Annals of Mathematics Studies

Number 142

Renormalization and 3-Manifolds which Fiber over the Circle

by

Curtis T. McMullen

PRINCETON UNIVERSITY PRESS

———

PRINCETON, NEW JERSEY

1996

The Annals of Mathematics Studies are edited by
Luis A. Caffarelli, John N. Mather, and Elias M. Stein

Princeton University Press books are printed on acid-free paper and meet the
guidelines for permanence and durability of the Committee on Production
Guidelines for Book Longevity of the Council on Library Resources

Printed in the United States of America by Princeton Academic Press

10 9 8 7 6 5 4 3 2 1

Library of Congress Cataloging-in-Publication Data
McMullen, Curtis T.
Renormalization and 3-manifolds which fiber over the circle / by Curtis T.
McMullen.
p. cm. — (Annals of mathematics studies : 142)
Includes bibliographical references and index.
ISBN 0-691-01154-0 (cl : alk. paper). — ISBN 0-691-01153-2 (pb : alk. paper)
1. Three-manifolds (Topology) 2. Differentiable dynamical systems.
I. Title. II. Series: Annals of mathematics studies : no. 142.
QA613. M42 1996
514'.3—dc20   96-19081

The publisher would like to acknowledge the author of this volume for
providing the camera-ready copy from which this book was printed

Contents

Renormalization and 3-Manifolds which
Fiber over the Circle

1 Introduction

In the late 1970s Thurston constructed hyperbolic metrics on most 3-manifolds which fiber over the circle. Around the same time, Feigenbaum discovered universal properties of period doubling, and offered an explanation in terms of renormalization. Recently Sullivan established the convergence of renormalization for real quadratic mappings.

In this work we present a parallel approach to renormalization and to the geometrization of 3-manifolds which fiber over the circle. This analogy extends the dictionary between rational maps and Kleinian groups; some of the new entries are included in Table 1.1.

Dictionary	
Kleinian group $\Gamma \cong \pi_1(S)$	Quadratic-like map $f : U \to V$
Limit set $\Lambda(\Gamma)$	Julia set $J(f)$
Bers slice B_Y	Mandelbrot set M
Mapping class $\psi : S \to S$	Kneading permutation
$\psi : AH(S) \to AH(S)$	Renormalization operator R_p
Cusps in ∂B_Y	Parabolic bifurcations in M
Totally degenerate group Γ	Infinitely renormalizable polynomial $f(z) = z^2 + c$
Ending lamination	Tuning invariant
Fixed point of ψ	Fixed point of R_p
Hyperbolic structure on $M^3 \to S^1$	Solution to Cvitanović-Feigenbaum equation $f^p(z) = \alpha^{-1} f(\alpha z)$

Table 1.1.

Both discussions revolve around the construction of a nonlinear dynamical system which is conformally self-similar.

For 3-manifolds the dynamical system is a surface group acting conformally on the sphere via a representation

$$\rho : \pi_1(S) \rightarrow \mathrm{Aut}(\widehat{\mathbb{C}}).$$

Given a homeomorphism $\psi : S \rightarrow S$, we seek a discrete faithful representation satisfying

$$\rho \circ \psi_*^{-1}(\gamma) = \alpha \rho(\gamma) \alpha^{-1}$$

for some $\alpha \in \mathrm{Aut}(\widehat{\mathbb{C}})$. Such a ρ is a fixed point for the action of ψ on conjugacy classes of representations. This fixed point gives a hyperbolic structure on the 3-manifold

$$T_\psi = S \times [0,1]/(s,0) \sim (\psi(s),1)$$

which fibers over the circle with monodromy ψ. Indeed, the conformal automorphisms of the sphere prolong to isometries of hyperbolic space $\mathbb{H}^3$, and T_ψ is homeomorphic to $\mathbb{H}^3/\Gamma$ where Γ is the group generated by α and the image of ρ.

For renormalization the sought-after dynamical system is a degree two holomorphic branched covering $F : U \rightarrow V$ between disks $U \subset V \subset \mathbb{C}$, satisfying the Cvitanović-Feigenbaum functional equation

$$F^p(z) = \alpha^{-1} F(\alpha z)$$

for some $\alpha \in \mathbb{C}^*$. The *renormalization operator* $\mathcal{R}_p$ replaces F by its pth iterate F^p, suitably restricted and rescaled, and F is a fixed point of this operator.

In many families of dynamical systems, such as the quadratic polynomials $z^2 + c$, one sees cascades of bifurcations converging to a map $f(z) = z^2 + c^\infty$ with the same combinatorics as a fixed point of renormalization. In Chapter 9 we will show that $\mathcal{R}_p^n(f)$ converges *exponentially fast* to the fixed point F. Because of this convergence, quantitative features of F are reflected in f, and are therefore universal among all mappings with the same topology.

Harmonic analysis on hyperbolic 3-space plays a central role in demonstrating the attracting behavior of $\mathcal{R}_p$ and ψ at their fixed points, and more generally yields *inflexibility* results for hyperbolic 3-manifolds and holomorphic dynamical systems.

We now turn to a more detailed summary.

Hyperbolic manifolds. By Mostow rigidity, a closed hyperbolic 3-manifold is determined up to isometry by its homotopy type. An open manifold $M = \mathbb{H}^3/\Gamma$ with injectivity radius bounded above and below in the convex core can generally be deformed. However, such an M is naturally bounded by a surface ∂M with a conformal structure, and the shape of ∂M determines M up to isometry.

In Chapter 2 we show these open manifolds with injectivity bounds, while not rigid, are inflexible: a change in the conformal structure on ∂M has an exponentially small effect on the geometry deep in the convex core (§2.4). This inflexibility is also manifest on the sphere at infinity $\widehat{\mathbb{C}}$: a quasiconformal conjugacy from $\Gamma = \pi_1(M)$ to another Kleinian group Γ' is differentiable at certain points in the limit set Λ. These *deep points* $x \in \Lambda$ have the property that the limit set is nearly dense in small balls about x — more precisely, Λ comes with distance $r^{1+\epsilon}$ of every point in $B(x, r)$.

Chapter 3 presents a variant of Thurston's construction of hyperbolic 3-manifolds that fiber over the circle. Let $\psi : S \to S$ be a pseudo-Anosov homeomorphism of a closed surface of genus $g \geq 2$. Then the mapping torus T_ψ is hyperbolic. To construct the hyperbolic metric on T_ψ, we use a two-step iterative process.

First, pick a pair of Riemann surfaces X and Y in the Teichmüller space of S. Construct the sequence of quasifuchsian manifolds $Q(\psi^{-n}(X), Y)$, ranging in a Bers slice of the representation space of $\pi_1(S)$. Let $M = \lim Q(\psi^{-n}(X), Y)$. The Kleinian group representing $\pi_1(M)$ is *totally degenerate* — its limit set is a dendrite.

For the second step, iterate the action of ψ on the space of representations of $\pi_1(S)$, starting with M. The manifolds $\psi^n(M)$ are all isometric; they differ only in the choice of isomorphism between $\pi_1(M)$ and $\pi_1(S)$. A fundamental result of Thurston's — the double limit theorem — provides an algebraically convergent subsequence $\psi^n(M) \to M_\psi$. The theory of pleated surfaces gives an upper bound on the injectivity radius of M in its convex core. Therefore any *geometric* limit N of $\psi^n(M)$ is rigid, and so ψ is realized by an isometry $\alpha : M_\psi \to M_\psi$, completing the construction.

Mostow rigidity implies the full sequence converges to M_ψ. From the the inflexibility theory of Chapter 2, we obtain the sharper statement that $\psi^n(M) \to M_\psi$ exponentially fast.

The case of torus orbifold bundles over the circle, previously considered by Jørgensen, is discussed in §3.7. We also give an explicit example of a totally degenerate group with no cusps (see Figure 3.4).

Renormalization. The simplest dynamical systems with critical points are the quadratic polynomials $f(z) = z^2 + c$. In contrast to Kleinian groups, the consideration of limits quickly leads one to mappings not defined on the whole sphere.

A *quadratic-like map* $g : U \to V$ is a proper degree two holomorphic map between disks in the complex plane, with $\overline{U}$ a compact subset of V. Its *filled Julia set* is $K(g) = \bigcap g^{-n}(V)$.

If the restriction of an iterate $f^n : U_n \to V_n$ to a neighborhood of the critical point $z = 0$ is quadratic-like with connected filled Julia set, the mapping f^n is *renormalizable*. When infinitely many such n exist, we say f is *infinitely renormalizable*.

Basic results on quadratic-like maps and renormalization are presented in Chapter 4. In Chapter 5 we define *towers* of quadratic like maps, to capture geometric limits of renormalization. A tower

$$\mathcal{T} = \langle f_s : U_s \to V_s \ : \ s \in S \rangle$$

is a collection of quadratic-like maps with connected Julia sets, indexed by levels $s > 0$. We require that $1 \in S$, and that for any $s, t \in S$ with $s < t$, the ratio t/s is an integer and f_s is a renormalization of $f_s^{t/s}$.

A tower has *bounded combinatorics* and *definite moduli* if t/s is bounded for adjacent levels and the annuli $V_s - \overline{U_s}$ are uniformly thick. In Chapter 6 we prove the Tower Rigidity Theorem: a bi-infinite tower $\mathcal{T}$ with bounded combinatorics and definite moduli admits no quasiconformal deformations. (This result is a dynamical analogue of the rigidity of totally degenerate groups.)

To put this rigidity in perspective, note that a single quadratic-like map $f_1 : U_1 \to V_1$ is never rigid; an invariant complex structure for f_1 can be specified at will in the fundamental domain $V_1 - \overline{U_1}$. In a tower with $\inf S = 0$, on the other hand, f_1 is embedded deep within the dynamics of f_s for s near zero (since $f_s^{1/s} = f_1$). The rigidity of towers makes precise the intuition that a high renormalization of a quadratic-like map should be nearly canonical.

Chapter 7 presents a two-step process to construct fixed points of renormalization. The procedure is analogous to that used to find a

fixed-point of ψ. For renormalization, the initial data is a real number
c such that the critical point of $z \mapsto z^2 + c$ is periodic with period p.
The first step is to construct an infinite sequence of superstable points
c^{*n} in the Mandelbrot set by iterating the *tuning map* $x \mapsto c * x$. The
iterated tunings c^{*n} converge to a point c^∞ such that $f(z) = z^2 + c^\infty$
is infinitely renormalizable. In the classical Feigenbaum example,
$c = -1$ and c^{*n} gives the cascade of period doublings converging to
the Feigenbaum polynomial $f(z) = z^2 + (-1)^\infty = z^2 - 1.4101155 \cdots$.
The maps f and $\mathcal{R}_p(f)$ are quasiconformally conjugate near their
Julia sets.

The second step is to iterate the renormalization operator $\mathcal{R}_p$,
starting with the point f. By Sullivan's *a priori* bounds, we can pass
to a subsequence such that $\mathcal{R}_p^n(f)$ converges to a quadratic-like map
F. This F can be embedded in a tower

$$\mathcal{T} = \langle f_s : s \in S = \{ \dots, p^{-2}, p^{-1}, 1, p, p^2, \dots \} \rangle,$$

such that $f_1 = F$ and $f_{p^k} = \lim \mathcal{R}_p^{n+k}(f)$. In the limit we also have
a quasiconformal mapping $\phi : \mathbb{C} \to \mathbb{C}$ conjugating f_{ps} to f_s for each
s. By the Tower Rigidity Theorem, ϕ is a conformal map (in fact
$\phi(z) = \alpha z$), and thus F is a fixed point of renormalization. Just as
for 3-manifolds, we use rigidity of the *geometric limit* $\mathcal{T}$ to conclude
the dynamics is self-similar.

Much of the construction also works when c is complex, but some
steps at present require c real.

Deep points and uniform twisting. Chapters 8 and 9 develop
results leading to the proof that renormalization converges exponentially fast.

Chapter 8 exploits the tower theory further to study the geometry and dynamics of infinitely renormalizable maps $f(z) = z^2 + c$.
Assume f has bounded combinatorics and definite moduli. Then we
show:

1. The complement of the postcritical Cantor set $P(f)$ is a Riemann surface with bounded geometry.

2. The critical point $z = 0$ is a deep point of the Julia set $J(f)$. In
   particular, blowups of $J(f)$ about $z = 0$ converge to the whole
   plane in the Hausdorff topology.

3. There are small Julia sets everywhere in $J(f)$. More precisely, for any $z \in J(f)$ and $0 < r < 1$, there is a quadratic-like map

$$g = f^{-i} \circ f^j : U \to V$$

in the dynamics generated by f, such that $\operatorname{diam} J(g) \asymp r$ and $d(z, J(g)) = O(r)$.

Chapter 9 lays the foundations for a general theory of holomorphic dynamical systems $\mathcal{F}$ and their geometric limits. With these foundations in place, it is possible to prove inflexibility results generalizing those for hyperbolic manifolds. The notion of bounded injectivity radius is replaced by that of *uniform twisting*.

Roughly speaking, $(\mathcal{F}, \Lambda)$ is uniformly twisting if the geometric limits of $\mathcal{F}$ as seen from within the convex hull of Λ in $\mathbb{H}^3$ are very nonlinear. Condition (3) above implies that $(\mathcal{F}(f), J(f))$ is uniformly twisting, where $\mathcal{F}(f)$ is the full dynamics generated by f and f^{-1}.

The Deep Conformality Theorem asserts that a quasiconformal conjugacy between uniformly twisting systems is $C^{1+\alpha}$-conformal at the deep points of Λ. By (2) above, we conclude that any conjugacy from f to another quadratic-like map is differentiable at the critical point $z = 0$. Exponentially fast convergence of renormalization then follows.

To state the final result, we remark that a fixed point of renormalization F has a canonical maximal analytic continuation $\tilde{F} : W \to \mathbb{C}$. The domain W of $\tilde{F}$ is an open, dense subset of the plane. Then there is a $\lambda < 1$ such that for any compact $K \subset W$,

$$\sup_K |\tilde{F}(z) - \mathcal{R}_p^n(f)(z)| < \lambda^n$$

for all f with the same inner class as F, and all $n \gg 0$.

The fixed point constructions for mapping classes and for renormalization operators are compared retrospectively in Chapter 10. We conclude with some open problems; among them, the conjectural self-similarity of the boundary of Teichmüller space, as observed in computer experiments conducted jointly with Dave Wright. These parallels and open questions are summarized in Table 1.2.

Harmonic analysis. Two appendices develop the analytic foundations of our results. Appendix A is devoted to quasiconformal flows

Theorems and Conjectures	
Compactness of $\psi^n(M)$ (Thurston's double limit theorem)	Compactness of $\mathcal{R}_p^n(f)$ (Sullivan's *a priori* bounds)
Rigidity of double limits	Rigidity of towers
Compactness + rigidity $\Longrightarrow$ convergence	Compactness + rigidity $\Longrightarrow$ convergence
Exponential convergence of $\psi^n(M)$	Exponential convergence of $\mathcal{R}_p^n(f)$
Deep points in Λ	Deep points in J
Injectivity radius bounded above in the convex core	Quadratic-like dynamics at every scale
Λ is locally connected	J is locally connected
area$(\Lambda) = 0$	area$(J) = 0$?
dim$(\Lambda) = 2$	dim$(J) = 2$?
Geodesic flow ergodic	Tower dynamics ergodic?
Bers' boundary self-similar?	Mandelbrot set self-similar?
Ending lamination conjecture	Mandelbrot set locally connected?

Table 1.2.

and Reimann's theorem, that a vector field with strain in L^∞ generates a unique quasiconformal isotopy. The proof is streamlined by showing a function with first derivatives in BMO satisfies a Zygmund condition.

Appendix B is devoted to the visual extension of deformations from S_∞^{n-1} to $\mathbb{H}^n$. This extension has been studied by Ahlfors, Reimann, Thurston and others.

A key role for us is played by the *visual distortion* $Mv(p)$ of a vector field v on S_∞^{n-1}, as seen from $p \in \mathbb{H}^n$. By definition $Mv(p)$ is the minimum, over all conformal vector fields w, of the maximum visual length of $(v-w)$ as seen from p. The quantity Mv depends only on the strain Sv (the Beltrami differential $\mu = \bar\partial v$ when $n = 3$), but in a subtle way. Our inflexibility results are all proved by bounding Mv.

An illustration. The intuitive link between rigidity, deep points in the Julia set and convergence of renormalization is the following. Let f be the Feigenbaum polynomial, and let ϕ be a quasiconformal conjugacy between f and $f \circ f$. The dilatation of ϕ is specified by a field of infinitesimal ellipses supported outside the Julia set $J(f)$ and invariant under the action of f. To visualize a typical invariant ellipse field, first consider a family of ellipses in the plane of constant eccentricity whose major axes are along rays through the origin. This ellipse field is invariant under the mapping $z \mapsto z^2$. Now the Riemann mapping from the outside of the unit disk to the outside of the Feigenbaum Julia set transports the dynamics of z^2 to that of f. Since the Julia set is quite dense near the postcritical set, the argument of the derivative of the Riemann mapping varies wildly, and so the resulting ellipse field is more or less random. The stretching in different directions approximately cancels out, so that ϕ, while fluctuating at very small scales, is close to a conformal map near the critical point. The Julia set of a high renormalization of f also resides in a small neighborhood of the critical point. Since ϕ also conjugates f^{2^n} to $f^{2^{n+1}}$, these two mappings are nearly conformally conjugate, and in the limit we obtain a fixed point of renormalization.

A blowup of the Feigenbaum Julia set near the critical point appears in Figure 1.3. The tree-like black regions are the points outside the Julia set. The thin postcritical Cantor set, lying on the real axis and evidently well-shielded from the ellipse field, is also

Figure 1.3. Asymptotic rigidity near the postcritical set.

shown in black. The white region is a 1-pixel neighborhood of the Julia set itself (which is nowhere dense).

Notes and references. Thurston's construction of hyperbolic structures on 3-manifolds which fiber over the circle appears in [Th5]; an early account is in [Sul1]. Recently Otal has given a self-contained presentation of Thurston's theorem, using the method of ℝ-trees to establish the double limit theorem [Otal].

Feigenbaum's work appears in [Feig]; similar discoveries were made independently by Coullet and Tresser [CoTr]. Many additional contributions to the theory of renormalization are collected in [Cvi]. Sullivan's *a priori* bounds and a proof of convergence of renormalization using Riemann surface laminations appear in [Sul5] and [MeSt].

The dictionary between rational maps and Kleinian groups was introduced in [Sul4]. See [Mc5] for an illustrated account, centering on classification problems for conformal dynamical systems.

A brief discussion of the results presented here and other iterations on Teichmüller space appeared in [Mc3]. This work is a sequel to [Mc4], which discussed the foundations of renormalization, and to [Mc2], which began the program of using analytic estimates in an "effective" deformation theory for Kleinian groups.

Parts of this work were presented at CUNY and IHES in 1989 and 1990, at the Boston University Geometry Institute in 1991 and in the Keeler Lectures at University of Michigan in 1993. A preliminary version of this manuscript was circulated in Fall of 1994. Many useful corrections and suggestions were provided by G. Anderson and the referee. This research was funded in part by the Miller Institute for Basic Research and the NSF. I would like to thank all for their support.

2   Rigidity of hyperbolic manifolds

This chapter begins with basic facts about complete hyperbolic manifolds and their geometric limits. We then give a proof of rigidity for manifolds whose injectivity radius is bounded above. Mostow rigidity for closed manifolds is a special case; the more general result will be used in the construction of hyperbolic manifolds which fiber over the circle.

The proof of rigidity combines geometric limits with the Lebesgue density theorem and the a.e. differentiability of quasiconformal mappings. This well-known argument is carried further in §2.4 to show certain open hyperbolic manifolds, while not rigid, are *inflexible* — any deformation is asymptotically isometric in the convex core. This inflexibility is also manifest on the sphere at infinity: quasiconformal conjugacies are automatically *differentiable* and *conformal* at certain points in the limit set. These results will be applied to surface groups in Chapter 3.

Useful references on hyperbolic geometry include [CEG], [BP], [Th1] and [Th4].

2.1   The Hausdorff topology

We will frequently need to make precise the idea that certain manifolds, dynamical systems, compact sets or other objects converge *geometrically*. To formulate this notion, we recall the Hausdorff topology.

Let X be a separable metric space, and let $\mathrm{Cl}(X)$ be the space of all closed subsets of X. The *Hausdorff topology* on $\mathrm{Cl}(X)$ is defined by saying $F_i \to F$ if

> (a) every neighborhood of a point $x \in F$ meets all but finitely many F_i; and
>
> (b) if every neighborhood of x meets infinitely many F_i, then $x \in F$.

A subset of $\mathrm{Cl}(X)$ is *closed* if it contains all its sequential limits.

Given an arbitrary sequence $\langle F_i \rangle$ of closed sets in X, we define $\liminf F_i$ as the largest set satisfying condition (a), and $\limsup F_i$ as the smallest set satisfying condition (b). Both $\liminf F_i$ and $\limsup F_i$ are closed and $\liminf F_i \subset \limsup F_i$. We have $F_i \to F$ if and only if $\liminf F_i = \limsup F_i = F$.

Proposition 2.1 *The space* $\mathrm{Cl}(X)$ *is sequentially compact in the Hausdorff topology.*

Proof [Haus, §28.2]: Let F_i be a sequence in $\mathrm{Cl}(X)$, and let U_k be a countable base for the topology on X. For each k, if $U_k \cap F_i \neq \emptyset$ for infinitely many i, then we may pass to a subsequence such that $U_k \cap F_i \neq \emptyset$ for all but finitely many i. Diagonalizing, we obtain a subsequence F_{i_n} which converges. Indeed, if $x \in \limsup F_{i_n}$, then for any neighborhood U of x, $F_{i_n} \cap U \neq \emptyset$ for infinitely many n. But then F_{i_n} meets U for all but finitely many n, and thus $x \in \liminf F_{i_n}$. Since the upper and lower limits agree, the sequence converges.

■

Now suppose X is also locally compact. By separability, X can be exhausted by a countable sequence of compact sets, so its one-point compactification $X^* = X \cup \{\infty\}$ is metrizable. For each closed set $F \subset X$, let $F^* = F \cup \{\infty\} \subset X^*$, and define

$$\delta(F_1, F_2) \quad = \quad \inf\{\epsilon > 0 \; : \; F_1^* \text{ is contained in an } \epsilon\text{-neighborhood of } F_2^*, \text{ and vice-versa }\}.$$

Convergence in this metric is the same as Hausdorff convergence on $\mathrm{Cl}(X)$, so we have:

Corollary 2.2 *If* X *is a separable, locally compact metric space, then* $\mathrm{Cl}(X)$ *is a compact metric space.*

References: [Haus], [HY, §2-16], [Nad].

2.2 Manifolds and geometric limits

Definitions. *Hyperbolic space* $\mathbb{H}^n$ is a complete simply-connected n-manifold of constant curvature -1; it is unique up to isometry.

The *Poincaré ball* gives a model for hyperbolic space as the unit ball in $\mathbb{R}^n$ with the metric

$$ds^2 = \frac{4\,dx^2}{(1 - r^2)^2}.$$

The boundary of the Poincaré ball models the *sphere at infinity* S_∞^{n-1} for hyperbolic space, and the isometries of $\mathbb{H}^n$ prolong to conformal maps on the boundary.

In dimension three, the sphere at infinity can be identified with the Riemann sphere $\widehat{\mathbb{C}}$, providing an isomorphism between the orientation preserving group $\mathrm{Isom}^+(\mathbb{H}^n)$ and the group of fractional linear transformations $\mathrm{Aut}(\widehat{\mathbb{C}}) \cong PSL_2(\mathbb{C})$.

A *Kleinian group* Γ is a discrete subgroup of $\mathrm{Isom}(\mathbb{H}^n)$. A Kleinian group is *elementary* if it contains an abelian subgroup of finite index.

A *hyperbolic n-manifold* M is a complete Riemannian manifold of constant curvature -1. Any such manifold can be presented as a quotient $M = \mathbb{H}^n/\Gamma$ of hyperbolic space by a Kleinian group.

Orientation. All hyperbolic manifolds we will consider, including $\mathbb{H}^n$ itself, will be assumed *oriented*. The identification between $\mathbb{H}^n$ and the universal cover of M will be chosen to preserve orientation. Then the group $\Gamma = \pi_1(M)$ is contained in $\mathrm{Isom}^+(\mathbb{H}^n)$ and it is determined by M up to conjugacy.

The thick-thin decomposition. The *injectivity radius* of a hyperbolic manifold M at a point x is half the length of the shortest essential loop through x.

The Margulis Lemma asserts that a discrete subgroup of $\mathrm{Isom}(\mathbb{H}^n)$ generated by elements sufficiently close to the identity contains an abelian subgroup of finite index [BP, §D], [Th4, §4.1]. This result controls the geometry of the *thin part* $M_{(0,\epsilon]}$ of a hyperbolic manifold, i.e. the subset where the injectivity radius is less than ϵ. There is an $\epsilon_n > 0$ such that every component L of $M_{(0,\epsilon_n]}$ is either a collar neighborhood of a short geodesic, or a *cusp*, homeomorphic to $N \times [0,\infty)$ for some complete Euclidean $(n-1)$-manifold. In the universal cover $\mathbb{H}^n$, each component L of the thin part is covered by either an r-neighborhood of a geodesic, or by a horoball.

The *limit set* $\Lambda \subset S_\infty^{n-1}$ of a Kleinian group Γ is the set of accumulation points of Γx for any $x \in \mathbb{H}^n$; it is independent of x.

For $E \subset S_\infty^{n-1}$, the *convex hull* of E (denoted hull(E)) is the smallest convex subset of $\mathbb{H}^n$ containing all geodesics with both endpoints in E. The *convex core* K of a hyperbolic manifold $M = \mathbb{H}^n/\Gamma$ is given by $K = \text{hull}(\Lambda)/\Gamma$. The convex core supports the recurrent part of the geodesic flow; it can also be defined as the closure of the set of closed geodesics. We say M is *geometrically finite* if a unit neighborhood of its convex core has finite volume.

The open manifold M can be prolonged to a *Kleinian manifold*

$$\overline{M} = (\mathbb{H}^n \cup \Omega)/\Gamma,$$

where $\Omega = S_\infty^{n-1} - \Lambda$ is the *domain of discontinuity* of Γ. In dimension $n = 3$, Ω can be identified with a domain on the Riemann sphere on which Γ acts holomorphically, so

$$\partial M = \Omega/\Gamma$$

carries the structure of a complex one-manifold (possibly disconnected).

To pin Γ down precisely, one may choose a frame ω over a point $p \in M$; then there is a unique Γ such that the standard frame at the origin in the Poincaré ball lies over the chosen frame ω on M. Conversely, any discrete torsion-free group $\Gamma \subset \text{Isom}^+(\mathbb{H}^n)$ determines a manifold with baseframe (M, ω) by taking $M = \mathbb{H}^n/\Gamma$ and $\omega =$ the image of the standard frame at the origin. When we speak of properties of M holding *at* the baseframe ω, we mean such properties hold at the point p over which the baseframe lies.

Geometric limits. The *geometric topology* on the space of hyperbolic manifolds with baseframes is defined by $(M_i, \omega_i) \to (M, \omega)$ if the corresponding Kleinian groups converge in the Hausdorff topology on closed subsets of $\text{Isom}(\mathbb{H}^n)$. In this topology, the space of all hyperbolic manifolds (M, ω) with injectivity radius greater than $r > 0$ at the baseframe ω is compact.

Here is a more intrinsic description of geometric convergence: $(M_i, \omega_i) \to (M, \omega)$ if and only if, for each compact submanifold $K \subset M$ containing the baseframe ω, there are smooth embeddings $f_i : K \to M_i$, defined for all i sufficiently large, such that f_i sends ω to ω_i and f_i tends to an isometry in the C^∞ topology. The last condition can be made precise by passing to the universal cover: then we obtain mappings $\widetilde{f}_i : \widetilde{K} \to \mathbb{H}^n$, sending the standard baseframe at the origin to itself; and we require that f_i tends to the identity mapping in the topology of C^∞-convergence on compact subsets of $\widetilde{K}$. (See [BP, Thm. E.1.13].)

When viewed from deep in the convex hull, the limit set of any hyperbolic manifold is nearly dense on the sphere at infinity. Here is a precise statement:

Proposition 2.3 *Let (M_i, ω_i) be a sequence of hyperbolic n-manifolds with baseframes in their convex cores. Suppose the distance from ω_i to the boundary of the convex core of M_i tends to infinity.*

Then the corresponding limit sets Λ_i converge to S_∞^{n-1} in the Hausdorff topology.

Proof. Let $B_i \subset S_\infty^{n-1} - \Lambda_i$ be a spherical ball of maximum radius avoiding the limit set. The circle bounding B_i extends to a hyperplane H_i in $\mathbb{H}^n$ bounding a half-space outside the convex hull of the limit set. Since the origin of the Poincaré ball corresponds to the baseframe ω_i, the hyperbolic distance from the origin to H_i is tending to infinity. But this means the spherical radius of B_i is tending to zero, so for i large the limit set comes close to every point on the sphere.

$\blacksquare$

If $\Gamma_i \to \Gamma$ is a geometrically convergent sequence of Kleinian groups, then

$$\Lambda(\Gamma) \subset \liminf \Lambda(\Gamma_i).$$

This follows from that fact that repelling fixed points of elements of Γ are dense in its limit set. However, the limit set can definitely shrink in the limit. For example, the Fuchsian groups

$$\Gamma(p) = \{\gamma \in PSL_2(\mathbb{Z}) : \gamma \equiv I \bmod p\}$$

converge geometrically to the trivial group (with empty limit set) as $p \to \infty$, even though $\Lambda(\Gamma(p)) = S_\infty^1$ for all p (since $\mathbb{H}^2/\Gamma(p)$ has finite volume).

The situation is more controlled if the injectivity radius is bounded *above*. Given $R > r > 0$, let $\mathcal{H}_{r,R}^n$ denote the space of all hyperbolic n-manifolds (M, ω) such that:

1. the baseframe ω is in the convex core of M;

2. the injectivity radius of M is greater than r at ω; and

3. the injectivity radius is bounded above by R throughout the convex core of M.

Proposition 2.4 *The space $\mathcal{H}^n_{r,R}$ is compact in the geometric topology, and the limit set varies continuously on this space.*

Proof. First let $M = \mathbb{H}^n/\Gamma$ be any hyperbolic manifold. In terms of the universal cover, the injectivity radius at a point x is given by

$$r(x, \Gamma) \;=\; \frac{1}{2} \inf_{\gamma \in \Gamma,\, \gamma \neq \mathrm{id}} d(x, \gamma x).$$

Let

$$T(\Gamma, R) = \{x \in \mathbb{H}^n \;:\; r(x, \Gamma) \leq R\}.$$

We claim

$$\overline{T(\Gamma, R)} \cap S^{n-1}_\infty \subset \Lambda.$$

Indeed, a point $y \in M$ where the injectivity radius is less than R lies on an essential loop of length at most $2R$; shrinking this loop, we find y lies within a distance D (depending only on R) of either a closed geodesic or a component of the Margulis thin part of M. Lifts of the closed geodesic lie in the convex hull of the limit set; lifts of the thin part touch the sphere at infinity in the limit set and have Euclidean diameters tending to zero. In either case, we conclude that a point $y \in T(\Gamma, R)$ which is close to the sphere at infinity is also close to the limit set, and the claim follows.

Now let (M_i, ω_i) be a sequence in $\mathcal{H}^n_{r,R}$; by the lower bound on the injectivity radius, we can assume the sequence converges geometrically to some based manifold (M, ω). Let $\Gamma_i \to \Gamma$ be the corresponding sequence of Kleinian groups.

For any D and $x \in \mathbb{H}^n$, the set of hyperbolic isometries with $d(x, \gamma x) \leq D$ is compact, so we have $r(x, \Gamma_i) \to r(x, \Gamma)$ uniformly on compact subsets of $\mathbb{H}^n$. Therefore

$$\limsup T(\Gamma_i, R) \subset T(\Gamma, R)$$

with respect to the Hausdorff topology on closed subsets of $\mathbb{H}^n$. By hypothesis, the injectivity radius is bounded above by R in the convex core of M_i, so hull(Λ_i) is contained in $T(\Gamma_i, R)$. Therefore

$$\limsup \text{hull}(\Lambda_i) \subset T(\Gamma, R).$$

Since the origin lies in hull(Λ_i) for all i, $T(\Gamma, R)$ contains all limits of rays from the origin to Λ_i, and thus

$$\limsup \Lambda(\Gamma_i) \subset \overline{T(\Gamma, R)} \cap S_\infty^{n-1} \subset \Lambda(\Gamma).$$

The inclusion $\Lambda(\Gamma) \subset \liminf \Lambda(\Gamma_i)$ holds generally, so we have shown convergence of the limit sets. It is easy to verify that $(M, \omega) \in \mathcal{H}^n_{r,R}$.

■

A similar result appears in [KT].

2.3 Rigidity

In this section we discuss various notions of rigidity for hyperbolic manifolds, and prove a rigidity theorem for manifolds with upper bounds on their injectivity radii. This result will suffice for later applications to 3-manifolds which fiber over the circle.

Definitions. A diffeomorphism $f : X \to Y$ between Riemannian n-manifolds is an *L-quasi-isometry* if

$$\frac{1}{L} \leq \frac{|Df(v)|}{|v|} \leq L$$

for every nonzero tangent vector v to X.

A homeomorphism $\phi : X \to Y$ (for $n > 1$) is *K-quasiconformal* if ϕ has distributional first derivatives locally in L^n, and

$$\frac{1}{K} |\det D\phi(x)| \leq \left(\frac{|D\phi(v)|}{|v|} \right)^n \leq K |\det D\phi(x)|$$

for almost every x and every nonzero vector $v \in \mathrm{T}_x X$ (see §A.2).

A hyperbolic manifold $M = \mathbb{H}^n/\Gamma$ is *quasiconformally rigid* if any quasiconformal map $\phi : S_\infty^{n-1} \to S_\infty^{n-1}$, conjugating Γ to another

Kleinian group Γ', is conformal. Similarly M is *quasi-isometrically rigid* if any quasi-isometry $f : M \to M'$, where M' is also hyperbolic, is homotopic to an isometry.

The following result is well-known.

Theorem 2.5 *Let $f : M_1 \to M_2$ be a κ-quasi-isometry between hyperbolic n-manifolds $M_i = \mathbb{H}^n/\Gamma_i$, $i = 1, 2$. Then the lift*

$$\tilde{f} : \mathbb{H}^n \to \mathbb{H}^n$$

of f to the universal covers extends continuously to a $K(\kappa)$-quasiconformal map on S_∞^{n-1} conjugating Γ_1 to Γ_2. The constant $K(\kappa)$ tends to one as $\kappa \to 1$.

Conversely, in dimension three, a K-quasiconformal mapping

$$\phi : S_\infty^2 \to S_\infty^2$$

conjugating Γ_1 to Γ_2 has a continuous extension to an equivariant $\kappa(K)$-quasi-isometry

$$\tilde{f} : \mathbb{H}^3 \to \mathbb{H}^3$$

which descends to a quasi-isometry $f : M_1 \to M_2$. The constant $\kappa(K)$ tends to one as $K \to 1$.

The quasiconformality of the boundary values of quasi-isometries is a key step in the proof of Mostow rigidity, and is true under weaker hypotheses ($\tilde{f}$ need only distort large distances by a bounded factor); see [Mos], [Th1, §5]. For the converse in dimension three, see Corollary B.23.

Corollary 2.6 *If M is a quasiconformally rigid hyperbolic n-manifold, then it is quasi-isometrically rigid.*

Proof. Given a quasi-isometry $f : M \to M'$, the lifted map $\tilde{f} : \mathbb{H}^n \to \mathbb{H}^n$ extends to a quasiconformal conjugacy between Γ and Γ'; by hypothesis the boundary mapping is actually *conformal*, so it agrees with the boundary values of an equivariant isometry $\tilde{\alpha}$ of $\mathbb{H}^n$. A homotopy can be constructed by interpolating along the geodesic joining $\tilde{f}(x)$ to $\tilde{\alpha}(x)$ for each x.

■

Invariant line fields. We now identify the sphere at infinity S^2_∞ with the Riemann sphere $\widehat{\mathbb{C}}$. Let $L^1(\widehat{\mathbb{C}}, dz^2)$ denote the Banach space of measurable integrable quadratic differentials $\psi = \psi(z)dz^2$ on the sphere, with the norm

$$\|\psi\| = \int_{\widehat{\mathbb{C}}} |\psi(z)| \, |dz|^2.$$

The absolute value of a quadratic differential is an area form, so the norm above is conformally natural; that is, for any $A \in \mathrm{Aut}(\widehat{\mathbb{C}})$, $\|A^*\phi\| = \|\phi\|$.

The dual of $L^1(\widehat{\mathbb{C}}, dz^2)$ is $M(\widehat{\mathbb{C}}) = L^\infty(\widehat{\mathbb{C}}, d\bar{z}/dz)$, the Banach space of bounded measurable Beltrami differentials $\mu(z) \, d\bar{z}/dz$ equipped with the sup-norm. The pairing between $L^1(\widehat{\mathbb{C}}, dz^2)$ and $L^\infty(\widehat{\mathbb{C}}, d\bar{z}/dz)$ is given by

$$\langle \mu, \psi \rangle = \int_{\widehat{\mathbb{C}}} \mu(z)\psi(z) \, |dz|^2;$$

it is also conformally natural.

The *weak* topology* on $M(\widehat{\mathbb{C}})$ is defined by $\mu_n \to \mu$ if and only if

$$\langle \mu_n, \psi \rangle \to \langle \mu, \psi \rangle$$

for every $\psi \in L^1(\widehat{\mathbb{C}}, dz^2)$.

A *line field* is a Beltrami differential with $|\mu| = 1$ on a set E of positive measure and $|\mu| = 0$ elsewhere. The tangent vectors ξ such that $\mu(\xi) = 1$ span a measurable field of tangent lines over E, and such a line field determines μ.

Proposition 2.7 *A hyperbolic 3-manifold $M = \mathbb{H}^3/\Gamma$ is quasiconformally rigid if and only if there is no Γ-invariant measurable line field on the sphere at infinity.*

Proof. The complex dilatation $\mu = \phi_{\bar{z}}/\phi_z$ of a quasiconformal conjugacy between a pair of Kleinian groups Γ and Γ' is a Γ-invariant Beltrami differential; if ϕ is not conformal, then $\mu/|\mu|$ provides an Γ-invariant line field on the set of positive measure where $\mu \neq 0$. Conversely, if μ is an invariant line field, then for any complex t with $|t| < 1$ there is a quasiconformal mapping ϕ_t with dilatation $t\mu$ [AB], and ϕ_t conjugates Γ to another Kleinian group Γ'. (Conceptually, $t\mu$ is a new complex structure invariant under Γ; ϕ_t provides a change of coordinates transporting $t\mu$ to the standard structure on the sphere.)

∎

The unit ball in $M(\widehat{\mathbb{C}})$ is compact in the weak* topology. Thus any sequence of line fields μ_n has a weak*-convergent subsequence. However, the weak* limit need not be a line field: for example, the limit μ may equal 0 if μ_n is highly oscillatory. In any case we still have:

Proposition 2.8 *If $\Gamma_n \to \Gamma$ geometrically and μ_n are invariant line fields for Γ_n, then any weak* limit μ of μ_n is Γ-invariant.*

Proof. For any $\psi \in L^1(\widehat{\mathbb{C}}, dz^2)$, $\gamma \in \Gamma$ and $\epsilon > 0$, there is a neighborhood U of γ in $\mathrm{Aut}(\widehat{\mathbb{C}})$ such that $\|\delta_*\psi - \gamma_*\psi\| < \epsilon$ for all δ in U. By the definition of geometric convergence, there is a $\gamma_n \in \Gamma_n \cap U$ for all n sufficiently large, and thus:

$$|\langle \mu_n - \gamma^*\mu_n, \, \psi \rangle| = |\langle \mu_n, \, \psi - \gamma_*\psi \rangle| \leq$$
$$\epsilon + |\langle \mu_n, \, \psi - (\gamma_n)_*\psi \rangle| = \epsilon + |\langle \mu_n - \gamma_n^*\mu_n, \, \psi \rangle| = \epsilon,$$

where we have used the fact that $\|\mu_n\| = 1$. From the definition of weak* limit it follows that $\langle \gamma^*\mu, \psi \rangle = \langle \mu, \psi \rangle$ for every ψ, and thus μ is Γ-invariant.

∎

A group Γ is quasiconformally rigid *on its limit set* if there is no Γ-invariant line field supported on Λ. This means any quasiconformal conjugacy ϕ which is conformal outside Λ is a Möbius transformation.

A line field is *parabolic* if it is given by $\mu = A^*(d\bar{z}/dz)$ for some $A \in \mathrm{Aut}(\widehat{\mathbb{C}})$. A parabolic line field is tangent to the pencil of circles passing through a given point in a given direction; see Figure 2.1. When $\mu = d\bar{z}/dz$, the circles become the horizontal lines in the plane.

We now show that any hyperbolic manifold whose injectivity radius is bounded above is quasiconformally rigid. More generally we have:

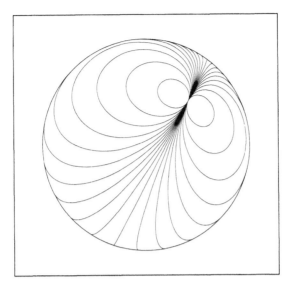

Figure 2.1. A parabolic line field on the sphere.

Theorem 2.9 (Bounded rigidity) *A hyperbolic 3-manifold $M = \mathbb{H}^3/\Gamma$ whose injectivity radius is bounded above throughout its convex core is quasiconformally rigid on its limit set.*

Proof. Suppose to the contrary that Γ admits an invariant line field μ on its limit set. By the Lebesgue density theorem, there is a point $p \in \mathbb{C}$ where $|\mu(p)| = 1$ and μ is almost continuous; that is, for any $\epsilon > 0$,
$$\lim_{r \to 0} \frac{\text{area}\{z \in B(p,r) \; : \; |\mu(z) - \mu(p)| < \epsilon\}}{\text{area } B(p,r)} = 1.$$
Here $\mu = \mu(z)d\bar{z}/dz$, and area and distance are measured in the Euclidean metric.

After conjugating by a Möbius transformation, we can assume that $p = 0$ and that $z = \infty$ is also in the limit set Λ of Γ. This implies that the geodesic g from $z = 0$ to $z = \infty$ lies in the convex hull of the limit set.

Orient g towards $z = 0$, and consider its image in M. Either g returns infinitely often to the thick part of M, or g enters a component of the thin part of M and never exits. In the latter case,

$z = 0$ must be the fixed point of some nontrivial element of Γ (a hyperbolic element for a short geodesic, or a parabolic element for a cusp). Since Γ is countable, we can choose p so these cases are avoided, and therefore g recurs infinitely often to the thick part of M.

Thus we have an $r > 0$ and points $x_n \in g$ tending to $z = 0$ such that the injectivity radius of M at $[x_n] \in \mathbb{H}^3/\Gamma$ is at least r. Let $\Gamma_n = A_n^{-1}\Gamma A_n$, where $A_n(z) = a_n z$ maps the origin of the Poincaré ball to x_n; then $a_n \to 0$ and we can assume $a_n > 0$. The group Γ_n leaves invariant the line field $\mu_n = A_n^*(\mu)$, and μ_n converges weak* to the parabolic line field $\mu_\infty = \mu(0)d\bar{z}/dz$ as n tends to infinity.

By construction, Γ_n belongs to $\mathcal{H}_{r,R}^3$ where R is an upper bound on the injectivity radius in the convex core of M. Thus we can pass to a subsequence such that $\Gamma_n \to \Gamma_\infty \in \mathcal{H}_{r,R}^3$. The limiting group Γ_∞ leaves the parabolic line field μ_∞ invariant.

Since $z = 0$ is a point of Lebesgue density of the limit set $\Lambda(\Gamma)$, the magnified limit sets

$$\frac{1}{a_n}\Lambda(\Gamma) \;=\; \Lambda(\Gamma_n)$$

converge to the whole sphere in the Hausdorff topology. Thus the limit set of Γ_∞ is also the whole sphere (by Proposition 2.4).

But any automorphism of the sphere preserving μ_∞ must fix the point at infinity, contradicting the fact that every orbit of Γ_∞ on the sphere is dense.

∎

As is well-known, rigidity fails in dimension 2: even a closed surface can admit many distinct hyperbolic structures.

On the other hand, the proof just given generalizes easily to hyperbolic n-manifolds, $n \geq 3$. If $\phi : S_\infty^{n-1} \to S_\infty^{n-1}$ is a quasiconformal conjugacy between Γ and Γ', then ϕ is differentiable almost everywhere. The pullback $\phi^*\sigma$ of the spherical metric determines an ellipsoid in the tangent space to almost every point. The vectors maximizing the ratio $(\phi^*\sigma)(v)/\sigma(v)$ span a canonical subspace $E_x \subset T_x S_\infty^{n-1}$, which cuts the ellipsoid in a round sphere of maximum radius. (On the 2-sphere, E_x is just the line field of major axes of the ellipses, or the whole tangent space at points where $D\phi$ is conformal.)

If $D\phi$ is not conformal a.e. on the limit set, then there is a set of positive measure $F \subset \Lambda$ over which rank E_x is a constant k with $0 < k < n - 1$. Then $E_x|F$ is a Γ-invariant k-plane field. Given an upper bound on the injectivity radius of the convex core of $M = \mathbb{H}^n/\Gamma$, we can blow up a point of almost continuity of $E_x|F$, pass to a geometric limit and obtain a contradiction as in the case of line fields. In summary we have:

Theorem 2.10 *Let $M = \mathbb{H}^n/\Gamma$ be a hyperbolic manifold of dimension $n \geq 3$ whose injectivity radius is bounded above throughout its convex core. Then Γ admits no measurable invariant k-plane field on its limit set, $0 < k < n - 1$, and M is quasiconformally rigid.*

Note that Γ need not be finitely generated.

Ergodicity versus rigidity. By a more subtle argument, Sullivan shows a Kleinian group admits no invariant k-plane field ($0 < k < n-1$) on the part of S_∞^{n-1} where its action is *conservative* [Sul3]. (The action of a discrete group on a measure space is conservative if there is no set A of positive measure such that the translates $\{\gamma(A) : \gamma \in \Gamma\}$ are disjoint.) It is easy to show that an upper bound on the injectivity radius in the convex core implies Γ acts conservatively on its limit set. Thus Sullivan's result implies the preceding Theorem.

On the other hand, there exist hyperbolic 3-manifolds $M = \mathbb{H}^3/\Gamma$ with bounded injectivity radius such that Γ does not act ergodically on the sphere. Thus ergodicity is stronger than rigidity.

For example, let M be the covering space of a closed hyperbolic 3-manifold N induced by a surjective mapping $\pi_1(N) \to \mathbb{Z} * \mathbb{Z}$ (see Figure 2.2). The injectivity radius of M is bounded above since the kernel of ρ is nontrivial (a closed manifold does not have the homotopy type of a bouquet of circles). Almost every geodesic on M wanders off to one of the ends of the free group, which form a Cantor set. Those landing in a given nonempty open set of ends have positive measure, and determine a Γ-invariant set of positive measure on the sphere. Since a Cantor set is the union of two disjoint nonempty open sets, the sphere is the union of two Γ-invariant sets of positive measure.

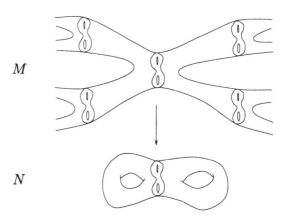

M

N

Figure 2.2. A $\mathbb{Z} * \mathbb{Z}$ covering space.

The origins of the line field viewpoint can be found in [Ah5] and [Sul3].

2.4   Geometric inflexibility

Let M be a hyperbolic 3-manifold whose injectivity radius is bounded above and below in its convex core, but with $\partial M \neq \emptyset$. Then M need not be rigid; deformations are often possible by changing the conformal structure on ∂M.

In this section we push the logic of geometric limits further to show a deformation of M decays exponentially fast within its convex core. In other words, the geometry of M deep within the core is inflexible: it changes only a small amount, even under a substantial deformation of ∂M. Our main result is:

Theorem 2.11 (Geometric inflexibility) *Let* $\Psi : M \to M'$ *be an L-quasi-isometry between a pair of hyperbolic 3-manifolds. Suppose the injectivity radius of M in its convex core K ranges in the interval $[R_0, R_1]$, where $R_0 > 0$.*

Then there is a volume-preserving quasi-isometry $\Phi : M \to M'$, *boundedly homotopic to* Ψ, *such that the pointwise quasi-isometry constant $L(\Phi, p)$ satisfies*

$$L(\Phi, p) \;\leq\; 1 + C \exp(-\alpha \, d(p, M - K)).$$

The constants C and $\alpha > 0$ depend only on (R_0, R_1, L).

Here a *bounded homotopy* is one which moves points a bounded hyperbolic distance. The maps Φ and Ψ are boundedly homotopic if and only if they admit lifts to $\mathbb{H}^3$ which agree on S^2_∞.

When $K = M$, the Theorem says Ψ is homotopic to an isometry. Thus Mostow rigidity for closed manifolds is a special case, and the Theorem can be thought of as an "effective" version of rigidity for open manifolds. These effective bounds are most interesting when M is geometrically infinite — then Φ is exponentially close to an isometry deep in the convex core. But the Theorem also has content when the convex core is compact, because the constants depend on M only via its injectivity radii.

Figure 2.3. The visual extension from ∂M to M.

Idea of the proof. A vector field v on S^{n-1}_∞ has a canonical *visual extension* to a vector field $V = \mathrm{ex}(v)$ on $\mathbb{H}^n$. When v represents a quasiconformal deformation of ∂M, V gives a quasi-isometric deformation of M. For $p \in M$, the metric distortion $SV(p)$ is the expected value of the quasiconformal distortion Sv at the endpoint of a random geodesic ray γ from p to ∂M (see Figure 2.3). Because of the injectivity bounds on M, a geodesic starting deep in the core tends to twist quite a bit before reaching ∂M, so under parallel transport the phase of the tensor Sv becomes almost random. Thus there is quite a bit of cancellation in the visual average, so the strain $SV(p)$ is small. This establishes inflexibility for a deformation, and the result for mappings follows in dimension three using the Beltrami equation.

The proof of geometric inflexibility will rely on properties of the conformal strain Sv, the visual extension $\mathrm{ex}(v)$ and the visual distortion Mv developed in Appendices A and B. We will assume familiarity with this material throughout this section.

Definitions. Let $M = \mathbb{H}^n/\Gamma$ be a hyperbolic manifold, $n \geq 3$. A *deformation* of Γ (or of M) is a vector field v on S^{n-1}_∞ such that $\gamma_*(v) - v$ is a conformal vector field for all $\gamma \in \Gamma$. A deformation is *trivial* if v is conformal. Two deformations v_1 and v_2 are *equivalent* if $v_1 - v_2$ is conformal.

A deformation can be thought of as an infinitesimal map conjugating Γ to another Kleinian group. The trivial deformations correspond to moving Γ by conjugacy inside $\mathrm{Isom}(\mathbb{H}^n)$.

A deformation is *quasiconformal* if v is a quasiconformal vector field; that is, if v is continuous and its conformal strain Sv is in L^∞ as a distribution. (On the Riemann sphere this condition is the same as $\|\bar{\partial}v\|_\infty < \infty$.) By considering the eigenspaces of the strain tensor Sv, we see a nontrivial quasiconformal deformation determines a Γ-invariant k-plane field on the sphere for some k, $0 < k < n-1$. Thus we have the infinitesimal form of Theorem 2.10:

Proposition 2.12 *Let $M = \mathbb{H}^n/\Gamma$ be a hyperbolic manifold whose injectivity radius is bounded above. Then any quasiconformal deformation of M is trivial.*

To explore the effect of a quasiconformal deformation v on the geometry of $M = \mathbb{H}^n/\Gamma$, let $V = \mathrm{ex}(v)$ be the *visual extension* of v to $\mathbb{H}^n$ (§B.1). Then $\gamma_*(V) - V$ is an infinitesimal isometry of $\mathbb{H}^n$ for each $\gamma \in \Gamma$. The extended vector field V is volume-preserving, so its conformal strain SV also measures its distortion of the hyperbolic metric. Since the strain of an isometric vector field is zero, the tensor SV is Γ-invariant, and therefore it descends to a strain field on M which we continue to denote by SV.

To illustrate the idea of inflexibility, we first show:

Proposition 2.13 *Suppose p lies in the convex core K of a hyperbolic n-manifold M whose injectivity radius is bounded above by R_1 on K and below by $R_0 > 0$ at p. Let v be a quasiconformal deformation of M, and let $V = \mathrm{ex}(v)$. Then*

$$\|SV(p)\| \leq \delta(d(p, \partial K)) \cdot \|Sv(p)\|_\infty$$

where $\delta(r) \to 0$ as $r \to \infty$, and $\delta(r)$ depends only on (n, R_0, R_1).

Proof. If not, we can find a sequence of hyperbolic manifolds $M_i = \mathbb{H}^n/\Gamma_i$ with points p_i in their convex cores K_i, and deformations v_i such that $\|SV_i(p)\| = 1$, $\|Sv_i\|_\infty$ is bounded by a constant k, and $d(p_i, \partial K_i) \to \infty$. (Here $V_i = \mathrm{ex}(v_i)$.) Lift to the universal cover so that $\tilde{p}_i = 0$ is the origin in the ball model for hyperbolic space; then $\|SV_i(0)\| = 1$.

Since $d(0, \partial \widetilde{K_i}) \to 0$, the limit set of Γ_i converges to the whole sphere in the Hausdorff topology. The injectivity radius of M_i is bounded above on K_i and below at p_i, so by Proposition 2.4 we can pass to a subsequence such that Γ_i converges geometrically to a Kleinian group Γ whose limit set is the whole sphere. The injectivity radius of $M = \mathbb{H}^n/\Gamma$ is bounded above by R_1, so M admits no quasiconformal deformations.

Now by Corollary A.11, the space of k-quasiconformal vector fields, modulo conformal vector fields, is compact in the topology of uniform convergence on the sphere. Thus after correcting by conformal vector fields (to obtain equivalent deformations), we may also assume v_i converges uniformly to a k-quasiconformal vector field v. Then v is a deformation of Γ, and v is nontrivial because $\|SV(0)\| = \lim \|SV_i(0)\| = 1$, where $V = \mathrm{ex}(v)$. This contradicts rigidity of Γ and establishes the Proposition.

■

The preceding Proposition bounds the strain deep in the core in terms of the strain on the sphere at infinity. The main step in the proof of inflexibility is the next Lemma, showing a bound on the distortion over a large *finite* sphere gives an improved bound at the center of the sphere. When iterated, this improvement yields exponential decay of deformations; and when integrated, the bounds for deformations give bounds for mappings.

The *visual distortion* $Mv : \mathbb{H}^n \to \mathbb{R}$ is defined by

$$Mv(p) = \inf_{Sw=0} \|v - w\|_\infty(p),$$

where $\|v - w\|_\infty(p)$ denotes the maximum length of the vector field $(v - w)$ as seen from p (§B.3). Like the strain SV, Mv only depends on v modulo conformal vector fields. Thus Mv is Γ-invariant and it too descends to a function on M.

Although we are mostly interested in bounding $\|SV(p)\|$, we will do so by first bounding the visual distortion. Since $\mu = Sv$ determines v up to a conformal vector field, Mv really only depends on μ — but in a rather implicit way. The use of the visual distortion Mv to measure the size of μ is crucial; the proof breaks down with many other natural measurements of the size of μ.

Let $S(p, r)$ denote the hyperbolic sphere of radius r centered at p.

Lemma 2.14 (Geometric decay) *Let v be a deformation of a hyperbolic manifold $M = \mathbb{H}^n/\Gamma$ with convex core K. Suppose the injectivity radius of M is bounded above by R_1 on K and below by $R_0 > 0$ at p. Then there is a radius $r(n, R_0, R_1)$ such that whenever $S(p, r) \subset K$, we have*

$$Mv(p) \;\leq\; \frac{1}{2} \sup_{q \in S(p,r)} Mv(q).$$

Proof. Again the proof is by contradiction. We will work in the universal cover, normalizing so $\tilde{p} = 0$. If the Lemma is false we can find a sequence of Kleinian groups Γ_i, a sequence of deformations v_i and a sequence of radii $r_i \to \infty$, such that $Mv_i(\tilde{p}) > 1/2$ but $Mv_i(q) \leq 1$ on the sphere $S(\tilde{p}, r_i) \subset \tilde{K}_i$. The bounds on the injectivity radius imply that after passing to a subsequence, Γ_i tends geometrically to a Kleinian group Γ whose limit set is the whole sphere, and the injectivity radius of $M = \mathbb{H}^n/\Gamma$ is bounded above by R_1.

After passing to a subsequence and correcting by conformal vector fields, we can assume v_i converges uniformly to a quasiconformal vector field v (see Corollary B.18). Since the convergence is uniform, $Mv(\tilde{p}) = \lim Mv_i(\tilde{p}) \geq 1/2$. But since M is rigid, v is conformal, and therefore $Mv = 0$. The Lemma follows by contradiction.

■

Next we show the visual extension of a deformation tends to an (infinitesimal) isometry exponentially fast in the convex core.

Remark on notation. Here and in the sequel, C_n and C'_n denote constants that depend only on the dimension n. Different occurrences of these constants are meant to be independent.

Theorem 2.15 (Infinitesimal inflexibility) *Let M be a hyperbolic n-manifold, $n \geq 3$. Suppose the injectivity radius of M in its convex core K ranges in the interval $[R_0, R_1]$, where $R_0 > 0$.*

Let $V = \mathrm{ex}(v)$ be the visual extension of a quasiconformal deformation v of M. Then for any $p \in M$ we have:

$$\|SV(p)\| \leq C_n M v(p) \leq C'_n \exp(-\alpha\, d(p, M - K)) \|Sv\|_\infty.$$

Here $\alpha > 0$ depends on (n, R_0, R_1).

Proof. By Theorem B.15, we have

$$\|SV(p)\| \leq C_n M v(p) \leq C'_n \|Sv\|_\infty.$$

Thus we need only establish the second inequality in the statement of the Theorem, and we may assume $p \in K$.

Let r be the radius guaranteed by Lemma 2.14 for the constants (R_0, R_1). Let N be the largest integer such that $d(p, \partial K) \geq Nr$. Then we can apply Lemma 2.14 N times to conclude that

$$Mv(p) \leq \frac{1}{2^N} \sup_{q \in S(p, Nr)} Mv(q).$$

Now $1/2^N \leq 2 \exp(-\alpha\, d(p, \partial K))$ where $\alpha = (\log 2)/r$, and $Mv(q) \leq C_n \|Sv\|_\infty$ by Theorem B.15, so the stated bound on $Mv(p)$ follows.

■

To give a global version of the preceding result, we need to show a point deep in the convex core remains reasonably deep after a quasi-isometry.

Proposition 2.16 *Let $\Phi : \mathbb{H}^n \to \mathbb{H}^n$ be an L-quasi-isometry, and let Λ be a closed subset of S^{n-1}_∞. Then $\Phi(\mathrm{hull}(\Lambda))$ is contained within a $d(n, L)$-neighborhood of $\mathrm{hull}(\Phi(\Lambda))$.*

Proof. Suppose the origin in the ball model for hyperbolic space is contained in the convex hull of Λ. Then there exist two points $x, x' \in \Lambda$ whose angular separation is at least $\pi/4$; otherwise Λ (and its convex hull) would be contained in a hemisphere. Thus any point $p \in \text{hull}(\Lambda)$ lies within a universally bounded distance d_0 of a geodesic $\gamma \subset \text{hull}(\Lambda)$. Now $\Phi(\gamma)$ is a quasi-geodesic with endpoints in $\Phi(\Lambda)$, so it lies within distance $d_1(n, L)$ of a geodesic $\gamma' \subset \text{hull}(\Phi(\Lambda))$ (see, e.g. [Th1, Prop. 5.9.2]), and therefore

$$d(\Phi(p), \text{hull}(\Phi(\Lambda))) \leq d(\Phi(p), \gamma') \leq Ld_0 + d_1(n, L).$$

■

Corollary 2.17 *For any* $p \in M$,

$$d(\Phi(p), M' - K') \;\geq\; \frac{d(p, M - K)}{L} - d(n, L).$$

Proof. We may assume $K' \neq M'$, since otherwise both sides are infinite. Suppose $\Phi(p) \in K'$. Let $\Phi(x)$ be the endpoint of the ray constructed by following the shortest geodesic from $\Phi(p)$ to $\partial K'$ and then continuing distance $d(n, L)$ further. Applying the preceding Proposition to Φ^{-1}, we find $x \notin \text{int } K$. Thus

$$\begin{aligned} d(p, M - K) &\leq d(p, x) \leq Ld(\Phi(p), \Phi(x)) \\ &\leq L(d(\Phi(p), M' - K') + d(n, L)). \end{aligned}$$

Solving for $d(\Phi(p), M' - K')$ gives the Corollary.

The argument when $\Phi(p) \notin K'$ is similar.

■

Proof of Theorem 2.11 (Geometric inflexibility). Let $M = \mathbb{H}^3/\Gamma$, let $M' = \mathbb{H}^3/\Gamma'$, and let $\tilde{\Psi} : \mathbb{H}^3 \to \mathbb{H}^3$ denote a lift of Ψ to the universal cover. Then the boundary values of Ψ give a $K(L)$-quasiconformal map $\psi : S^2_\infty \to S^2_\infty$ conjugating Γ to Γ'. Applying Theorem B.22, we can construct a Beltrami isotopy ϕ_t such that $\phi_0 = \text{id}$, $\phi_1 = \psi$, and ϕ_t conjugates Γ to a Kleinian group Γ_t. The Beltrami isotopy is the integral of a quasiconformal vector field v_t satisfying $\|\bar{\partial}v_t\| \leq k(L)$.

Now apply the visual extension to obtain a time-dependent vector field $V_t = \mathrm{ex}(v_t)$ on $\mathbb{H}^3$. The integral of this vector field gives a family of volume-preserving quasi-isometries $\Phi_t : \mathbb{H}^3 \to \mathbb{H}^3$ prolonging ϕ_t (by Theorem B.21). The quasi-isometry constant of Φ_t is bounded by a constant L' depending only L. This isotopy of $\mathbb{H}^3$ descends to a family of maps $M \to M_t = \mathbb{H}^3/\Gamma_t$ which we will also denote by Φ_t.

Since $\phi_1 = \psi$, $\Phi_1 : M \to M_1 = M'$ is homotopic to Ψ. To complete the proof, we will bound the quasi-isometry constant $L(\Phi_1, p)$.

Let K_t denote the convex core of M_t. By Theorem B.21, the quasi-isometry constant is bounded by the integral of the strain of V_t:

$$\log L(\Phi_1, p) \;\leq\; \int_0^1 \|SV_t(\Phi_t(p))\| \, dt. \tag{2.1}$$

Now it follows from our hypothesis on M and Proposition 2.16 that the injectivity radius of M_t in its convex core K_t lies in an interval $[R_0', R_1']$ depending only on $[R_0, R_1]$ and L. Applying Theorem 2.15, we have

$$\|SV_t(\Phi_t(x))\| \leq C' \exp(-\alpha' \, d(\Phi_t(x), M_t - K_t)) \, \|Sv_t\|_\infty, \tag{2.2}$$

where C' and α' also depend only on (R_0, R_1, L).

By the preceding Corollary, there is a constant $d(L')$ such that

$$d(\Phi_t(x), M_t - K_t) \;\geq\; \frac{d(x, M - K)}{L'} - d(L').$$

Using the fact that $\|Sv_t\|_\infty = \|\bar{\partial} v_t\|_\infty < k(L)$, we may rewrite equation (2.2) in the form

$$\|SV_t(\Phi_t(x))\| \;\leq\; C \exp(-\alpha \, d(x, M - K))$$

for suitable C and α. Combining this bound with equation (2.1) gives

$$\log L(\Phi_1, p) \;\leq\; C \exp(-\alpha \, d(p, M - K));$$

so taking $\Phi = \Phi_1$, we have established the Theorem.

∎

2.5   Deep points and differentiability

A quasiconformal conjugacy between a pair of Kleinian groups is often nowhere differentiable on the limit set.

In this section we will show a conjugacy is sometimes forced to be differentiable and conformal at many points. This conformality can be thought of as a remnant of Mostow rigidity when the limit set is not the whole sphere. It says the fine structure in the limit set is unchanged by a quasi-isometric deformation.

Definitions. Let $\Lambda \subset S_\infty^{n-1}$ be a compact set, and let $K \subset \mathbb{H}^n$ be its convex hull. We say $x \in \Lambda$ is a *deep point* of Λ if there is a geodesic ray

$$\gamma : [0, \infty) \to K,$$

parameterized by arclength and terminating at x, such that for some $\delta > 0$,

$$d(\gamma(s), \partial K) \geq \delta s > 0$$

for all s. In other words, the depth of γ inside the convex hull of Λ increases linearly with hyperbolic length. When quantitative precision is required we say x is a δ-*deep* point.

In terms of the sphere at infinity, a point $x \in \Lambda$ is deep if and only if the blowups of Λ about x converge exponentially fast to the sphere in the Hausdorff metric on compact sets. Equivalently, let $B(x, r)$ be the spherical ball of radius r about x, and let $s(r)$ denote the radius of the largest ball contained in $B(x, r) - \Lambda$. Then x is deep if and only if there is a $\beta > 0$ such that $s(r) \leq r^{1+\beta}$ for all r sufficiently small.

A homeomorphism $\phi : \widehat{\mathbb{C}} \to \widehat{\mathbb{C}}$ is $C^{1+\alpha}$-*conformal* at z if, after changing coordinates so z and $\phi(z)$ lie in $\mathbb{C}$, the complex derivative $\phi'(z)$ exists and

$$\phi(z + t) = \phi(z) + \phi'(z) \cdot t + O(|t|^{1+\alpha})$$

for all $t \in \mathbb{C}$ sufficiently small. We may now state:

Theorem 2.18 (Deep conformality) *Let $M = \mathbb{H}^3/\Gamma$ be a hyperbolic 3-manifold whose injectivity radius is bounded above and below in its convex core, and let ϕ be a quasiconformal conjugacy from Γ to another Kleinian group Γ'. Then ϕ is $C^{1+\alpha}$-conformal at every deep point in the limit set Λ.*

More precisely, if the injectivity radius in the core ranges in
$[R_0, R_1]$, ϕ *is K-quasiconformal and x is a δ-deep point, then ϕ is*
$C^{1+\alpha}$ *conformal at x, where $\alpha > 0$ depends only on (R_0, R_1, K, δ).*

Proof. The proof follows the same lines as the proof of geometric
inflexibility (Theorem 2.11).

We will work in the upper half-space model $\mathbb{H}^3 = \mathbb{C} \times \mathbb{R}_+$ with
coordinates (z, t). Let $\gamma(s) = (0, e^{-s})$ denote the geodesic ray starting at $(0, 1)$ at terminating at $z = 0$. Let K be the convex hull of
the limit set of Γ.

By a conformal change of coordinates, we can arrange that the
deep limit point x is at the origin $z = 0$, that $\gamma(0) \in K$, and that

$$d(\gamma(s), \partial K) \geq \delta s > 0$$

for all $s > 0$. By conjugating Γ', we can also arrange that ϕ fixes 0,
1 and ∞.

Next we embed ϕ in a Beltrami isotopy ϕ_t, fixing 0, 1 and ∞, with
$\phi_0 = $ id and $\phi_1 = \phi$, using Theorem B.22. The isotopy ϕ_t integrates
a continuous vector field v_t with $\|\bar{\partial} v_t\|_\infty \leq k$, where k depends only
on the dilatation of ϕ. Let $V_t = \text{ex}(v_t)$, and integrate V_t to obtain a
quasi-isometric isotopy Φ_t of $\mathbb{H}^3$ prolonging ϕ_t. Each mapping Φ_t is
an L-quasi-isometry, where L depends only on k.

The mapping $\phi_t \cup \Phi_t$ on $S^2_\infty \cup \mathbb{H}^3$ conjugates Γ to a Kleinian group
Γ_t. Let K_t denote the convex hull of the limit set Λ_t of Γ_t.

We claim that

$$d(\gamma(s), \partial K_t) \geq \delta' s - d$$

for all $s > 0$, where $d, \delta' > 0$ are independent of t. Indeed, since
Φ_t is an L-quasi-isometry, $R_t = \Phi_t(\gamma[0, \infty))$ is a uniformly quasi-
geodesic ray, starting near $\gamma(0)$ and terminating at $z = 0$. Thus R_t
is contained in a uniformly bounded neighborhood of R_0, and the
point $\Phi_t(\gamma(s'))$ closest to $\gamma(s)$ satisfies $s' > s/L - O(1)$. Applying
Corollary 2.17 to estimate the change in the convex hull gives

$$\begin{aligned}
d(\gamma(s), \partial K_t) &\geq d(\Phi_t(\gamma(s')), \partial K_t) - O(1) \\
&\geq \frac{d(\gamma(s'), \partial K)}{L} - O(1) \geq \frac{\delta s}{L^2} - O(1),
\end{aligned}$$

which is a bound of the required form.

Since Φ_t is a quasi-isometry, the given upper and lower bounds on the injectivity radius of M in its convex core provide similar bounds for $M_t = \mathbb{H}^3/\Gamma_t$. Combining Theorem 2.15 with the estimate on the distance to the convex hull boundary just obtained, we conclude that the visual distortion of v_t tends to zero exponentially fast along the geodesic ray γ. That is,

$$Mv_t(\gamma(s)) \leq Ce^{-\alpha s}$$

for some $C, \alpha > 0$ independent of t.

By the normalization of the Beltrami isotopy, we have $v_t(0) = v_t(1) = v_t(\infty) = 0$. Thus the exponentially decay of the visual distortion implies, by Theorem B.26, that v_t is $C^{1+\alpha}$ at the origin. More precisely,

$$|v_t(z) - v_t'(0)z| \leq C'|z|^{1+\alpha}$$

when $|z| \leq 1$, where C' is independent of t. Applying Theorem B.27, we conclude that ϕ_1 is $C^{1+\alpha}$-conformal at the origin, as claimed.

■

A generalization of this pointwise conformality result to other holomorphic dynamical systems appears in Theorem 9.15.

2.6   Shallow sets

This section develops further the theme of conformality at deep points, but in the absence of dynamics. Instead we assume $\phi : \widehat{\mathbb{C}} \to \widehat{\mathbb{C}}$ is a quasiconformal map that is *conformal* on an open set Ω. We will show that conformality persists at points $x \in \partial\Omega$ that are well-surrounded by Ω. This well-surroundedness is guaranteed when x is a deep point of $\overline{\Omega}$ and $\partial\Omega$ is *shallow*.

Some applications to Kleinian groups and iterated rational maps are given in examples.

Definitions. A closed set $\Lambda \subset S_\infty^{n-1}$ is *R-shallow* if its convex hull in $\mathbb{H}^n$ contains no ball of radius R. We say Λ is *shallow* if it is *R*-shallow for some $R > 0$. The terminology is suggested by the fact that a shallow set has no deep points.

It is easy to see the following are equivalent:

1. Λ is shallow.

2. There is an $\epsilon > 0$ such that any spherical ball $B(x,r)$, contains a ball $B(y, \epsilon r)$ disjoint from Λ.

3. There is no sequence $g_n \in \mathrm{Aut}(S_\infty^{n-1})$ such that $g_n(\Lambda) \to S_\infty^{n-1}$ in the Hausdorff topology on compact subsets of the sphere.

In this section we will prove:

Theorem 2.19 *Let* $\phi : S_\infty^2 \to S_\infty^2$ *be a* K-*quasiconformal map that is conformal on an open set* Ω. *Assume* $\partial\Omega$ *is* R-*shallow. Then:*

1. *there is a natural extension of* ϕ *to a diffeomorphism* $\Phi : \mathbb{H}^3 \to \mathbb{H}^3$ *whose quasi-isometry constant satisfies*

$$L(\Phi, p) \leq 1 + C\exp(-\alpha\, d(p, \partial K))$$

   for all $p \in K = \mathrm{hull}(\overline{\Omega})$; *and*

2. *for any* δ-*deep point* x *of* $\overline{\Omega}$, ϕ *is* $C^{1+\beta}$-*conformal at* x.

The constants C, α *and* β *depend only on* K, R *and* δ.

The proof depends on some facts about shallow sets.

Proposition 2.20 *If* $\Lambda \subset S_\infty^{n-1}$ *is* R-*shallow, then the spherical volume of an* r-*neighborhood of* Λ *satisfies*

$$\mathrm{vol}\,\mathcal{N}(\Lambda, r) \leq C_n r^\alpha$$

where $\alpha > 0$ *depends only on* R.

Proof. Express Λ as a union of two R-shallow sets, each contained in a hemisphere. It suffices to prove the Proposition for each piece. Restricting attention to one piece, we may assume Λ is contained the the unit cube with respect to stereographic projection; that is,

$$\Lambda \subset C = [-1, 1]^{n-1} \subset \mathbb{R}_\infty^{n-1} \subset S_\infty^{n-1}.$$

Since the Euclidean and spherical metrics are quasi-isometric on the unit cube, it is enough to show $\mathrm{vol}\,\mathcal{N}(\Lambda, r) = O(r^\alpha)$ in the Euclidean metric on $\mathbb{R}_\infty^{n-1}$.

For any cube $D \subset \mathbb{R}_\infty^{n-1}$, let D_k denote the collection of subcubes of D of side length $1/2^k$ times that of D. There are $2^{k(n-1)}$ such cubes.

Since Λ is shallow, there is a $k > 0$ such that for any cube D, Λ is disjoint from some cube in D_k. Thus $\Lambda \cap D$ can be covered by $2^{k(n-1)} - 1$ cubes chosen from D_k. Since $\Lambda \subset C$, by induction, Λ is covered by $(2^{k(n-1)} - 1)^j$ cubes chosen from C_{jk}. Thus for $r = 1/2^{jk}$, we have $\operatorname{vol} \mathcal{N}(\Lambda, r) = O(s^j)$, where $s = (2^{k(n-1)} - 1)/2^{k(n-1)} < 1$. Choose $\alpha > 0$ such that $1/2^{\alpha k} < s$; then we have $\operatorname{vol} \mathcal{N}(\Lambda, r) = O(r^\alpha)$, as required.

$\blacksquare$

Dimension. A shallow set is also small in the sense of dimension. Recall that the *box dimension* of a compact metric space (X, d) is defined by

$$\text{box-dim}(X) = \limsup_{r \to 0} \frac{\log N(X, r)}{\log(1/r)},$$

where $N(X, r)$ is the minimum number of r-balls required to cover X. The box dimension is the infimum of those $\delta \geq 0$ such that $N(X, r) = O(r^{-\delta})$. The *Hausdorff dimension* $\operatorname{H.dim}(X)$ is the infimum of those $\delta \geq 0$ such that X can be covered by balls $B(x_i, r_i)$ with $\sum r_i^\delta$ arbitrarily small. Clearly $\operatorname{H.dim}(X) \leq \text{box-dim}(X)$.

Since $N(\Lambda, r) = O(r^{1-n} \operatorname{vol} \mathcal{N}(\Lambda, r))$, we have:

Corollary 2.21 *If $\Lambda \subset S_\infty^{n-1}$ is shallow, then the Hausdorff and box dimensions of Λ are both strictly less than $(n - 1)$.*

Compare [Sal], where shallow sets are called porous.

We now turn to the proof of Theorem 2.19. Let $\Omega \subset S_\infty^{n-1}$ be an open set such that $\partial\Omega$ is shallow, and let K be its convex hull. The *visual measure* of a set $E \subset S_\infty^{n-1}$ as seen from $p \in \mathbb{H}^n$ is the probability that a random geodesic ray from p lands in E (cf. §B.3).

Proposition 2.22 *Let $\Omega \subset S_\infty^{n-1}$ be an open set such that $\partial\Omega$ is R-shallow. Then there is an $\alpha(n, R) > 0$ such that for all $p \in K = \operatorname{hull}(\overline{\Omega})$, we have*

(the visual measure of $(S_\infty^{n-1} - \Omega)$ as seen from p)

$$\leq C_n \exp(-\alpha \, d(p, \partial K)).$$

Proof. Normalize coordinates so $p = 0$ in the ball model for hyperbolic space. By the definition of the convex hull, every spherical ball $B(x, r)$ meets $\overline{\Omega}$, where $r = C_n \exp(-d(p, \partial K))$. Thus $(S^{n-1}_\infty - \Omega) \subset \mathcal{N}(\partial\Omega, r)$. But $\operatorname{vol} \mathcal{N}(\partial\Omega, r) = O(r^\alpha)$ by Proposition 2.20, so the same bound holds for the visual measure of the complement of Ω.

∎

The visual distortion $Mv(p)$ is bounded in terms of the nth root of the visual measure of the support of Sv, by Corollary B.20; this yields:

Corollary 2.23 *Let v be a quasiconformal vector field on S^{n-1}_∞, such that $Sv = 0$ on Ω. Then*

$$Mv(p) \leq C_n \exp(-\alpha\, d(p, \partial K)/n)\|Sv\|_\infty$$

for all $p \in K$.

Proof of Theorem 2.19. The Theorem follows from the decay estimate on quasiconformal deformations above, by the same reasoning as in the proofs of Theorems 2.11 and 2.18. Include ϕ in a Beltrami isotopy ϕ_t with $\phi_0 = \operatorname{id}$ and $\phi_1 = \phi$. Then use the visual extension to extend the isotopy to a quasi-isometric flow Φ_t on $\mathbb{H}^3$. The quasiconformal image of a shallow set is shallow, so the Corollary above applies to the vector fields $v_t = d\phi_t/dt$ to show Mv_t decays exponential fast in the convex core. The integral of these bounds with respect to t controls to quasi-isometry constant of Φ_1 and the conformality of ϕ_1 at deep points.

∎

Measurable deep points. Now let $\Omega \subset S^2_\infty$ be a measurable set and let δ be positive. A point x is a *measurable δ-deep point* for Ω if $\operatorname{area}(B(x, r) - \Omega) = O(r^{2+\delta})$. Equivalently, the visual measure of the complement of Ω tends to zero exponentially fast along a geodesic ray in $\mathbb{H}^3$ tending to x. Note that x need not belong to Ω. By Proposition 2.22 we have:

Proposition 2.24 *If x is a deep point of $\overline{\Omega}$, and $\partial\Omega$ is shallow, then x is a measurable deep point of Ω.*

It is known that $\mathrm{vol}(\phi(E)) = O(\mathrm{vol}(E)^{1/K})$ for a K-quasiconformal map [Ast], so measurable deep points are preserved (with a controlled change in the exponent δ). Thus the argument proving Theorem 2.19 shows more generally:

Theorem 2.25 (Boundary conformality) *Let $\phi : S^2_\infty \to S^2_\infty$ be a K-quasiconformal map that is conformal on a measurable set Ω. Let x be a measurable δ-deep point for Ω. Then there is an $\alpha(K, \delta)$ such that ϕ is $C^{1+\alpha}$-conformal at x.*

Examples.

1. Let $\phi : \mathbb{C} \to \mathbb{C}$ be a quasiconformal map that is conformal outside the strip $S = \{z : |\operatorname{Im} z| < 1\}$. Then ϕ is $C^{1+\alpha}$-conformal at infinity, because infinity is a deep point of $\widehat{\mathbb{C}} - S$. This is consistent with the fact that for $a, b > 0$, the map

$$\phi(x + iy) = \begin{cases} ax + iy & \text{if } y = 1 \\ bx + iy & \text{if } y = -1 \end{cases}$$

   extends to a quasiconformal map of the plane if and only if $a = b$.

2. Here is a criterion for a limit set to be shallow.

   Proposition 2.26 *Let $\Lambda \subset S^2_\infty$ be the limit set of a finitely generated nonelementary Kleinian group Γ. Then Λ is shallow if and only if Γ is geometrically finite, with no cusps of rank 2, and $\Lambda \neq S^2_\infty$.*

   Proof. If Γ is shallow, then every point in the convex core K of $M = \mathbb{H}^3/\Gamma$ is within a bounded distance of ∂K. Since ∂K has finite area (by Ahlfors' finiteness theorem), the thick part of K has finite volume and so Γ is geometrically finite. A rank 2 cusp would give an end of M entirely contained in K, so these are ruled out.

Conversely, suppose Γ is geometrically finite, with no cusps of rank 2, and $\Lambda \neq S_\infty^2$. Then $\partial K \neq \emptyset$, and the part of K outside a neighborhood of the cusps is compact, so these points are within a bounded distance of the boundary. In the part of K meeting a rank 1 cusp, the distance to ∂K tends to zero, so $d(p, \partial K)$ is bounded uniformly for $p \in K$ and Λ is shallow.

∎

3. Let X be a compact Riemann surface of genus $g \geq 2$. Let Γ be a geometrically finite group in the boundary of a Bers slice B_X (see §3.2). Then Γ has a finite number of rank 1 accidental parabolic cusps. The domain of discontinuity Ω of Γ has a unique invariant component $\Omega(X)$, and $\Omega(X)/\Gamma = X$.

Now let $\phi : X \to Y$ be a quasiconformal map. Then we can use ϕ to transport the complex structure of Y to X, and then lift it to $\Omega(X)$ to obtain a new Γ-invariant complex structure on the sphere. Solving the Beltrami equation, we obtain a group $\Gamma' \in \partial B_Y$, and a quasiconformal conjugacy $\tilde{\phi}$ from Γ to Γ', such that $\tilde{\phi} : \Omega(X) \to \Omega(Y)$ is a lift of ϕ, and $\tilde{\phi}$ is conformal on $\Omega - \Omega(X)$.

By the preceding Proposition, the limit set $\Lambda = \partial\Omega(X)$ of Γ is shallow. On the other hand, each cusp is a deep point of $\Omega - \Omega(X)$. By Theorem 2.19 we conclude *the conjugacy $\tilde{\phi}$ is $C^{1+\alpha}$-conformal at all rank 1 cusps of Γ*.

4. Let $f(z)$ be a rational map with a parabolic fixed point at $z = p$, and let $\Omega = \{z \; : \; d(f^n(z), p) \to 0\}$ be its basin of attraction. It is well-known that Ω contains a bouquet of finitely many petals attached at p, each tangent to its neighbors (see e.g. [CG, §II.5]). Thus *a parabolic fixed-point is a measurable deep point of its basin*. Now let $\phi : \hat{\mathbb{C}} \to \hat{\mathbb{C}}$ be a quasiconformal conjugacy from f to another rational map g. By Theorem 2.25, *if ϕ is conformal on Ω, then ϕ is $C^{1+\alpha}$-conformal at p*.

The same result holds with f and g replaced by germs of holomorphic maps with parabolic fixed-points.

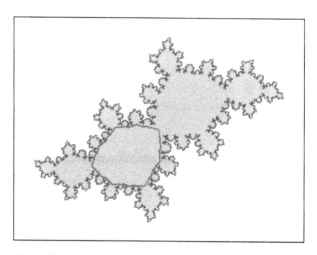

Figure 2.4. The filled Julia set of a quadratic polynomial with a golden mean Siegel disk.

5. Let $f(z) = e^{2\pi i\theta}z + z^2$, where θ is an irrational number of *bounded type*, meaning its continued fraction expansion

$$\theta = \cfrac{1}{a_1 + \cfrac{1}{a_2 + \cfrac{1}{a_3 + \cdots}}}$$

satisfies $\sup a_i < \infty$. Then $z = 0$ is the center of a Siegel disk D for $f(z)$. By a theorem of Herman and Świątek, ∂D is a quasicircle (see Figure 2.4; cf. [Her], [Dou2]). It can be shown that the Julia set $J(f)$ is shallow, and the critical point of f is a deep point of the filled Julia set $K(f)$. Now if $g : U \to V$ is a quadratic-like map with a fixed point having the same multiplier $e^{2\pi i\theta}$, then there exists a quasiconformal conjugacy ϕ from f to g near their filled Julia sets that is conformal on $\operatorname{int} K(f)$. By Theorem 2.19 we conclude *the conjugacy ϕ is $C^{1+\alpha}$ at the critical point of f.* Similar methods show the boundary of the Siegel disk is self-similar when θ is a quadratic irrational; see [Mc7] for more details.

3   Three-manifolds which fiber over the circle

This chapter presents the construction of hyperbolic 3-manifolds which fiber over the circle.

Consider a pseudo-Anosov homeomorphism $\psi : S \to S$ of a compact surface of negative Euler characteristic. To find a hyperbolic structure on the mapping torus T_ψ of ψ, one seeks a hyperbolic manifold M_ψ homeomorphic to $S \times \mathbb{R}$ on which the homotopy class ψ is represented by an isometry α. Then $M_\psi/\langle \alpha \rangle$ is homeomorphic to T_ψ.

It is easy to find hyperbolic structures on $S \times \mathbb{R}$ such that ψ is represented by a quasi-isometry. For example, a hyperbolic structure on S extends to $S \times \mathbb{R}$ in a unique way such that $S \times \{0\}$ is totally geodesic. This amounts to considering the action of a Fuchsian group on hyperbolic 3-space. Since ψ can be represented by a quasiconformal map on S, it can be represented by a quasi-isometry on $S \times \mathbb{R}$.

The first step in the construction is deform such a Fuchsian manifold to obtain a limit $M_{\psi,Y}$, with one geometrically finite end bounded by a Riemann surface Y, and one geometrically infinite end, still quasi-isometrically invariant by ψ.

The second step is to consider the sequence of manifolds $\psi^n(M_{\psi,Y})$. These manifolds differ only in the marking of their fundamental groups by $\pi_1(S)$. As $n \to \infty$, the generators of $\pi_1(S)$ are represented by geodesics deeper and deeper in the convex core of $M_{\psi,Y}$. By Thurston's double limit theorem, we can find an algebraic limit M_ψ. We extract a geometric limit N at the same time, and show it is rigid; then ψ is realized by an isometry on M_ψ, completing the construction.

Background on surface groups and marked hyperbolic manifolds is presented in the first two sections. In §3.3 we discuss the action of the mapping class group $\mathrm{Mod}(S)$ on Teichmüller space and on a Bers slice, and state the version of the double-limit theorem we will need. The hyperbolic structure on T_ψ is constructed in §3.4. In §3.5 a further analysis shows the manifold $M_{\psi,Y}$ is asymptotically isometric to M_ψ. When S is closed, this implies the injectivity radius of $M_{\psi,Y}$ is bounded above and below.

In 3.6 we use this information, and our earlier inflexibility results, to show $\psi^n(M_{\psi,Y})$ converges to M_ψ exponentially fast. This result sets the stage for a parallel development in the context of renormalization, realized in §9.5.

Finally in §3.7 we study the special case of torus orbifold bundles over the circle. An explicit example of a totally degenerate group is discussed; its limit set is reminiscent of the Feigenbaum Julia set, with elliptic points corresponding to the critical orbit.

3.1   Structures on surfaces and 3-manifolds

Let S be a compact connected oriented surface of negative Euler characteristic. We allow S to have boundary and we let $\mathrm{int}(S) = S - \partial S$.

The *Teichmüller space* $\mathrm{Teich}(S)$ classifies conformal structures on $\mathrm{int}(S)$ in which each boundary component corresponds to a puncture. A point in $\mathrm{Teich}(S)$ is specified by a Riemann surface X, each end of which is conformally isomorphic to a punctured disk, and a homeomorphism $f : \mathrm{int}(S) \to X$, sending the orientation on S to the canonical orientation of the complex manifold X. The map f is a *marking* of X.

Two marked surfaces (f_1, X_1) and (f_2, X_2) define the same point in Teichmüller space if there is a complex analytic isomorphism $\iota : X_1 \to X_2$ such that $\iota \circ f_1$ is homotopic to f_2. The *Teichmüller metric* is defined by

$$d(X_1, X_2) \;=\; \frac{1}{2}\inf \log K(\phi)$$

where $\phi : X_1 \to X_2$ ranges over all quasiconformal maps in the homotopy class $f_2 \circ f_1^{-1}$, and $K(\phi)$ denotes the dilatation of ϕ.

Although we will denote a typical point in $\mathrm{Teich}(S)$ by X, there is always an implicit marking $f : S \to X$.

We now consider 3-manifolds with the homotopy type of a surface. Let $H(S)$ be the set all marked hyperbolic 3-manifolds M with the same homotopy type as S. Here a marking of M is a homotopy equivalence $f : S \to M$, mapping ∂S into horoball neighborhoods of the cusps of M. Two marked hyperbolic manifolds (f_1, M_1) and (f_2, M_2) represent the same point in $H(S)$ iff there is an orientation-preserving isometry $\iota : M_1 \to M_2$ such that $\iota \circ f_1$ is homotopic to f_2.

Equivalently,

$$H(S) = D(S)/\operatorname{Isom}^+(\mathbb{H}^3),$$

where the isometry group acts by conjugation on the space $D(S)$ of all discrete faithful homomorphisms

$$\rho : \pi_1(S) \to \operatorname{Isom}^+(\mathbb{H}^3)$$

such that $\pi_1(B)$ maps to a parabolic subgroup for each component B of ∂S.

The *algebraic topology* on the set $H(S)$ is defined by $[\rho_n] \to [\rho]$ if there are representatives of each equivalence class such that $\rho_n(g) \to \rho(g)$ for each $g \in \pi_1(S)$. We let $AH(S)$ denote $H(S)$ endowed with the algebraic topology.

On the level of manifolds, $M_n \to M$ in $AH(S)$ iff there are smooth homotopy equivalences $h_n : M \to M_n$, compatible with markings, such that on any compact subset of M, h_n tends C^∞ to a local isometry for all $n \gg 0$.[1]

The *quasi-isometric topology* on $H(S)$ is defined by the metric

$$d((f_1, M_1), (f_2, M_2)) = \inf \log L(g)$$

where the infimum is over all quasi-isometric diffeomorphisms $g : M_1 \to M_2$ homotopic to $f_2 \circ f_1^{-1}$, and $L(g)$ is the quasi-isometry constant of g. If there is no quasi-isometry in the required homotopy class, the quasi-isometric distance is $+\infty$. The balls of finite radius form a basis for this topology; the resulting space is denoted $QH(S)$.

The mapping $QH(S) \to AH(S)$ is continuous. Using equicontinuity of suitably normalized quasi-isometries, one may easily check:

---

[1] It is *not* required that h_n converges to an embedding on compact sets; indeed, when the algebraic and geometric limits of M_n differ, h_n approximates a covering map rather than an embedding.

Proposition 3.1 *The quasi-isometric distance is lower semicontinuous on $AH(S) \times AH(S)$: if $A_i \to A$ and $B_i \to B$ in $AH(S)$, then*

$$d(A, B) \leq \liminf d(A_i, B_i).$$

Remark. In [Th2] and [Th5], $H(S)$ is denoted $H(S \times I, \partial S \times I)$. We have adopted the former notation for brevity, since we will almost exclusively be concerned with 3-manifolds homotopy equivalent to S.

3.2   Quasifuchsian groups

A *Fuchsian group* is a Kleinian group which stabilizes a round disk on the sphere at infinity. Any point in the Teichmüller space of S can be presented as a quotient $X = \mathbb{H}^2/\Gamma$ where $\mathbb{H}^2$ is the upper halfplane in $\mathbb{C}$ and Γ is a Fuchsian group. Considered as a Kleinian group, the domain of discontinuity Ω of Γ is the union of the upper and lower halfplanes.

A *quasifuchsian group* is a Kleinian group which is quasiconformally conjugate to a Fuchsian group. Let

$$QF(S) \subset H(S)$$

denote the subset of representations $[\rho : \pi_1(S) \to \text{Isom}^+(\mathbb{H}^3)]$ such that $\Gamma = \rho(\pi_1(S))$ is quasifuchsian. For any such group, the Kleinian manifold

$$\overline{M} = (\mathbb{H}^3 \cup \Omega)/\Gamma$$

is bounded by a pair of Riemann surfaces X and Y. There is a homeomorphism between $\overline{M}$ and $\text{int}(S) \times [0, 1]$, compatible with the marking of M and with orientation, that restricts to a marking of X by S and Y by $\overline{S}$ (the same surface with reversed orientation). Bers showed that the map

$$QF(S) \to \text{Teich}(S) \times \text{Teich}(\overline{S})$$

so determined is a bijection.

We will denote the inverse mapping by Q, and think of $Q(X, Y)$ as a hyperbolic 3-manifold determined by X and Y, equipped with an implicit homotopy equivalence $f : S \to Q(X, Y)$. The map Q

is a homeomorphism with respect to the product of the Teichmüller metrics on the domain and the algebraic *or* quasi-isometric topology on the range. The manifold $Q(X, Y)$ is Fuchsian if and only if Y is the complex conjugate of X.

A *Bers slice* $B_Y \subset AH(S)$ is the image of $\text{Teich}(S) \times \{Y\}$ under Q.

Proposition 3.2 *The closure of a Bers slice is compact in the algebraic topology.*

See [Bers1, Theorem 3], [Mc1, §6.3]; compactness comes from the fact the hyperbolic length of a geodesic in $Q(X, Y)$ is bounded in terms of the length of the corresponding geodesic on Y. Bers also showed the component $\Omega(Y)$ of the domain of discontinuity which uniformizes Y persists in the limit at all points in ∂B_Y.

A point $(f : S \to M)$ in the boundary of a Bers slice has an *accidental parabolic* if there is an essential loop γ on S, not homotopic into ∂S, such that $f(\gamma)$ represents a parabolic element in $\pi_1(M)$. A boundary point is *totally degenerate* if its domain of discontinuity consists of a single component, $\Omega(Y)$.

Proposition 3.3 (Bers) *A point in the boundary of a Bers slice which has no accidental parabolics is totally degenerate.*

See [Bers1, Prop. 7]. It follows that a generic boundary point (in the sense of Baire category) is totally degenerate, since the condition $\text{tr}(\rho(\gamma)) = \pm 2$ for an accidental parabolic determines a nowhere dense subset of the boundary.

Theorem 3.4 (Thurston, Bonahon) *For any hyperbolic manifold M in $AH(S)$, the injectivity radius of M in its convex core is bounded above by a constant R which depends only on S.*

Remarks. Logically we will only use the fact that this statement holds for certain groups in the boundary of a Bers slice. The proof of this case is in [Th1]; Thurston shows every point in the convex core of M is within a bounded distance of a *pleated surface*. Such a surface is a pair (f, X) consisting of a hyperbolic surface $X \in \text{Teich}(S)$ and a map $f : X \to M$, inducing an isomorphism on π_1, such that

through every point of X there is a geodesic segment which is mapped
isometrically to a geodesic segment in M. The injectivity radius of
X is bounded above in terms of S (because its area is determined
by Gauss-Bonnet), so we obtain an upper bound on the injectivity
radius in the convex core of M.

To find sufficiently many pleated surfaces, one interpolates between pairs of such surfaces by interpolating between their pleating
laminations. This idea completes the proof when M is quasifuchsian,
since the boundary faces of the convex core are pleated surfaces. In
general one must establish that the ends of the thick part of the
convex core of M are *geometrically tame*, meaning any neighborhood of the end contains a pleated surface. This is established for
$M \in \overline{QF(S)}$ in [Th1] and for all $M \in AH(S)$ in [Bon]. (It is conjectured that these two sets of manifolds are actually the same.) A
detailed proof of the Theorem above appears in [Can, Thm. 6.2].

The same discussion leads to a proof of:

Theorem 3.5 (Tameness) *Any M in $AH(S)$ is homeomorphic to*
$\mathrm{int}(S) \times \mathbb{R}$.

3.3   The mapping class group

The *mapping class group* $\mathrm{Mod}(S)$ is the group of isotopy classes
of orientation-preserving homeomorphisms from S to itself. It is
canonically isomorphic to the group of outer automorphisms of $\pi_1(S)$
that preserve orientation and stabilize the set of conjugacy classes of
boundary curves.

A mapping class $\psi : S \to S$ acts on $\mathrm{Teich}(S)$ by sending (f, X) to
$(f \circ \psi^{-1}, X)$. There are similar actions on $\mathrm{Teich}(\overline{S})$ and on $AH(S)$.
These actions are compatible with the mapping

$$Q : \mathrm{Teich}(S) \times \mathrm{Teich}(\overline{S}) \to AH(S);$$

that is, $\psi(Q(X,Y)) = Q(\psi(X), \psi(Y))$.

A mapping class also acts on a Bers slice B_Y by $Q(X,Y) \mapsto$
$Q(\psi(X), Y)$. The action of $\mathrm{Mod}(S)$ on B_Y is *not* the restriction of
its action on $AH(S)$; the latter sends B_Y to $B_{\psi(Y)}$.

A mapping class ψ is *reducible* if there is a finite system of disjoint
essential simple closed curves on S, none parallel to ∂S, which are

permuted up to isotopy.

Theorem 3.6 (Thurston, Bers) *A mapping class $\psi \in \mathrm{Mod}(S)$ either:*

1. *has finite order and fixes a point in $\mathrm{Teich}(S)$, or*

2. *is reducible and of infinite order, or*

3. *has no fixed point in $\mathrm{Teich}(S)$, but stabilizes a Teichmüller geodesic.*

See [FLP], [Th3], [Bers2]. In the last case ψ is *pseudo-Anosov*.

Here is a classification of surface homeomorphisms in terms of the dynamics of ψ acting on a Bers slice B_Y. For $M \in AH(S)$, we say ψ is *quasi-isometrically realized* on M if there is a quasi-isometry $\phi : M \to M$ such that $f \circ \psi$ is homotopic to $\phi \circ f$. Equivalently, the quasi-isometric distance $d(\psi(M), M)$ is finite.

Theorem 3.7 (Limits in a Bers slice) *Let $Q_0 = Q(X, Y)$ be any point in the Bers slice B_Y, and let $Q_\infty \in \overline{B_Y} \subset AH(S)$ be any accumulation point of the sequence $Q_i = Q(\psi^i(X), Y)$. Then either:*

1. *$Q_\infty \in B_Y$ is quasifuchsian and ψ has finite order, or*

2. *$Q_\infty \in \partial B_Y$ has an accidental parabolic, and ψ is reducible and of infinite order, or*

3. *$Q_\infty \in \partial B_Y$ is totally degenerate with no accidental parabolics, and ψ is pseudo-Anosov.*

In all cases, ψ is quasi-isometrically realized on Q_∞.

Remark. For applications to mapping tori we will only need the pseudo-Anosov case, which is essentially contained in [Bers3]. We will later see that $Q(\psi^i(X), Y)$ actually converges when ψ is pseudo-Anosov (Theorem 3.11). A refinement of the result above appears in [Sh].

Proof. The Teichmüller distance from $\psi^{i+1}(X)$ to $\psi^i(X)$ is bounded independent of i (since it is equal to the distance between $\psi(X)$ and

X). Thus for every i there is a uniformly quasiconformal conjugacy between $\psi(Q_i)$ and Q_i, and therefore

$$d(\psi(Q_i), Q_i) = O(1),$$

where $d(,)$ is the quasi-isometric distance. By lower semicontinuity of $d(,)$ on $AH(S) \times AH(S)$, the limiting manifold Q_∞ is also translated a bounded distance by ψ, so ψ is quasi-isometrically realized on Q_∞.

We now turn to the classification.

Since $\mathrm{Mod}(S)$ acts on Teichmüller space with discrete orbits and finite point stabilizers, Q_∞ lies in B_Y if and only if ψ has finite order.

Suppose $Q_\infty = \mathbb{H}^3/\Gamma_\infty$ has at least one accidental parabolic subgroup $\langle \gamma \rangle \subset \Gamma_\infty$; we will show ψ is reducible. Let g be the closed geodesic on Y corresponding to γ. There is a γ-invariant lift $\tilde{g}$ of g to $\Omega(Y)$, the component of the domain of discontinuity lying over Y. The geodesic $\tilde{g}$ converges at both ends to the fixed point of γ.

In the Poincaré metric on the disk $\Omega(Y)$, any two geodesics meet in at most one point. If $\tilde{h} \subset \Omega(Y)$ is another geodesic invariant by an accidental parabolic $\delta \notin \langle \gamma \rangle$, then $\tilde{g}$ and $\tilde{h}$ are disjoint — otherwise their closures would give two topological circles on the sphere crossing at one point.

Consequently the set of all accidental parabolics is represented by a system of disjoint simple closed geodesics $\mathcal{G}$ on Y. Their total number is finite (it is bounded by the topology of S). A quasi-isometry sends cusps to cusps, so ψ permutes $\mathcal{G}$ and is therefore reducible.

Finally suppose ψ is reducible and of infinite order; to complete the proof, we will show Q_∞ has an accidental parabolic.

For simplicity, assume there is a single simple nonperipheral loop g on S preserved by ψ up to isotopy (the case of a system of loops is similar). Pass to a subsequence such that $Q_i \to Q_\infty$ algebraically. Then the hyperbolic length of the geodesic representative of g in Q_i converges to a limit L. If $L = 0$ then g is an accidental parabolic in Q_∞ as desired.

Now suppose $L > 0$. By a result of Sullivan, the faces of the convex hull of Q_i are K-quasi-isometric to the hyperbolic surfaces $\psi^i(X)$ and Y with a universal constant K [Sul1], [EpM]. Since ψ stabilizes the isotopy class of g, the length of g on $\psi^i(X)$ is the same as its length on X, so it is bounded above independent of i. Thus

the shortest curve representing g on either face of the convex hull
has length bounded above. Since the length of g in Q_i is bounded
below, both faces are within a bounded distance of the geodesic representative of g.

In particular the two faces of the convex hull of Q_i are a bounded
distance apart. Equivalently, if we normalize the universal cover so
the lift of the face corresponding to Y passes through the center
of the hyperbolic ball, then $S_\infty^2 - \Omega(Y)$ contains a ball of definite
spherical radius. In the limit this implies the domain of discontinuity
of Q_∞ cannot consist solely of $\Omega(Y)$, and thus Q_∞ is not totally
degenerate. Since $Q_\infty \in \partial B_Y$, it must have an accidental parabolic
by Proposition 3.3.

∎

Double limits. Whereas a Bers slice is compact, the space $AH(S)$
is not. For example, the (noncompact) Teichmüller space of S is
properly embedded in $AH(S)$ as the space of Fuchsian groups. It
is thus tricky to guarantee compactness, let alone convergence, of
sequences in $AH(S)$.

Thurston's double limit theorem [Th5, Thm. 4.1] provides a compactness criterion in this setting. We state a simplified version that
suffices for applications to mapping tori.

Theorem 3.8 (Pseduo-Anosov double limits) *Let $\psi \in \mathrm{Mod}(S)$
be pseudo-Anosov. Then for any $(X, Y) \in \mathrm{Teich}(S) \times \mathrm{Teich}(\overline{S})$, the
quasifuchsian manifolds*

$$\{Q(\psi^{-i}(X), \psi^j(Y)) \;:\; i, j \geq 0\}$$

lie in a compact subset of $AH(S)$.

Remarks. In Thurston's compactification of Teichmüller space,
$\psi^{-i}(X)$ and $\psi^j(Y)$ converge to the projective classes of the stable and
unstable measured laminations (μ_s, μ_u) for ψ. These bind the surface in the sense that the intersection number $i(\nu, \mu_s) + i(\nu, \mu_u) > 0$
for any third measured lamination $\nu \neq 0$. Then compactness of
$Q(\psi^{-i}(X), \psi^j(Y))$ as $i, j \to \infty$ is immediate from the statement of
[Th5, Thm. 4.1].

If i or j stays bounded then there is a subsequence lying in a Bers slice, or a slice with the factors reversed; either slice has compact closure in $AH(S)$.

Otal has recently given a new proof of the double limit theorem, using $\mathbb{R}$-trees [Otal].

3.4   Hyperbolic structures on mapping tori

The *mapping torus* of a homeomorphism $\psi : S \to S$ is the oriented 3-manifold T_ψ obtained from the cylinder $S \times [0, 1]$ by identifying $(x, 0)$ with $(\psi(x), 1)$ for every $x \in S$. There is a natural fibration $T_\psi \to S^1$ with fiber S and monodromy ψ. The orientation of T_ψ is the product of the given orientation of S and the positive orientation of $[0, 1]$. We may now state:

Theorem 3.9 (Thurston) *The mapping torus T_ψ admits a complete hyperbolic structure of finite volume if and only if ψ is pseudo-Anosov.*

If ψ is reducible or of finite order, then $\pi_1(T_\psi)$ contains copies of $\mathbb{Z} \oplus \mathbb{Z}$ which do not correspond to the boundary, so T_ψ cannot be hyperbolic.

In this section we present a variant of Thurston's construction of a hyperbolic structure on T_ψ when ψ is pseudo-Anosov. The construction will be justified by assembling the previous results; it proceeds as follows.

1. First, pick a point $M_0 = (f_0 : S \to M)$ in the boundary of a Bers slice B_Y, such that $Q(\psi^{-i}(X), Y)$ accumulates at M_0 as $i \to \infty$. The manifold M_0 is totally degenerate, and it admits a quasi-isometry ϕ in the homotopy class of $\psi : S \to S$.

2. The sequence $M_n = \psi^n(M_0)$ lies in a compact subset of $AH(S)$ by Thurston's double limit theorem. Note that these M_n are simply different markings of the same manifold M.

3. Pass to a subsequence such that M_n converges algebraically to M_∞ and geometrically to N. There is a covering map $\pi : M_\infty \to N$. The manifold N has injectivity radius bounded above and its limit set is the whole sphere.

4. The quasi-isometry ϕ gives rise to limiting quasi-isometries ϕ_∞ and ξ such that the diagram

$$
\begin{array}{ccc}
M_\infty & \xrightarrow{\phi_\infty} & M_\infty \\
\pi \downarrow & & \pi \downarrow \\
N & \xrightarrow{\xi} & N
\end{array}
$$

commutes.

5. Since N is rigid, ξ is homotopic to an isometry of N; therefore ϕ_∞ is also homotopic to an isometry α of M_∞.

6. The mapping torus T_ψ is then homeomorphic to the finite volume manifold $M_\infty/\langle\alpha\rangle$.

We now fill in the details.

Proof of Theorem 3.9.

Step 1. Let (X, Y) be any pair of Riemann surfaces in $\mathrm{Teich}(S) \times \mathrm{Teich}(\overline{S})$. The quasifuchsian manifolds $Q(\psi^{-i}(X), Y)$ lie in a Bers slice B_Y, so as $i \to \infty$ they accumulate on some boundary point $M_0 = (f_0 : S \to M)$ in $AH(S)$. By Proposition 3.7, the limit is totally degenerate, and there is a quasi-isometry

$$
\phi : M \to M
$$

in the homotopy class of $\psi : S \to S$.

Step 2. The marked manifold

$$
M_n \;=\; \psi^n(M_0) \;=\; (f_0 \circ \psi^{-n} : S \to M) \;=\; (f_n : S \to M)
$$

is an accumulation point of $Q(\psi^{n-i}(X), \psi^n(Y))$ as $i \to \infty$. Thus the sequence M_n is precompact by Theorem 3.8 (Pseudo-Anosov Double Limits).

Step 3. Pass to a subsequence such that $(f_n : S \to M)$ converges algebraically to $(f_\infty : S \to M_\infty)$. Pick a baseframe ω_∞ in the convex core of M_∞, and consider a compact connected submanifold K containing the baseframe and the image $f_\infty(S)$. By algebraic convergence, there are local quasi-isometries $g_n : K \to M$, such that

$g_n \circ f_\infty$ is homotopic to f_n, and the quasi-isometry constant of g_n tends to one.

We can assume g_n is an isometry at the baseframe, and let ω_n be the image of ω_∞. Since

$$g_n \circ f_\infty : S \to M$$

is a homotopy equivalence, g_n cannot map K entirely into the thin part of M; thus the injectivity radius of M is bounded below at ω_n. Therefore after passing to a further subsequence, (M, ω_n) converges geometrically to a based manifold (N, ω).

We now verify properties of the geometric limit N. Choose K large enough that it contains a closed geodesic γ_∞ on M_∞, representing the homotopy class $f_\infty(\gamma)$ of some loop γ on S. Then

$$\gamma_n \;=\; g_n(\gamma_\infty)$$

represents the homotopy class $f_0 \circ \psi^{-n}(\gamma)$ on M, and its length is bounded above. Since ψ is pseudo-Anosov, these homotopy classes are all distinct as n varies, so γ_n must eventually exit any compact subset of M. Thus the distance from γ_n to the boundary of the convex core of M tends to infinity.

But the distance from γ_n to the baseframe ω_n is bounded by the diameter of $g_n(K)$, so the baseframe ω_n is also very deep in the convex core when n is large. Thus the limit set $\Lambda(M, \omega_n)$ converges to the whole sphere as $n \to \infty$ (Proposition 2.3). The injectivity radius of M in its convex core is bounded above by Theorem 3.4. By Proposition 2.4, the injectivity radius of N is also bounded above, and the limit set $\Lambda(N, \omega)$ is the whole sphere, being the Hausdorff limit of $\Lambda(M, \omega_n)$.

A limit of g_n gives a local isometry of K into N; letting K exhaust M_∞, we obtain a covering map $\pi : M_\infty \to N$.

Step 4. Identify the universal covers of (M, ω_n), $(M_\infty, \omega_\infty)$ and (N, ω) with $\mathbb{H}^3$ so the standard frame at the center of the Poincaré ball lies over the given frame on each manifold. The markings f_n and f_∞ give representations $\rho_n : \pi_1(S) \to \Gamma_n$ and $\rho_\infty : \pi_1(S) \to \Gamma_\infty$. The geometric limit of Γ_n is a Kleinian group $\Pi \supset \Gamma_\infty$ corresponding to N.

Let $\Psi : \pi_1(S) \to \pi_1(S)$ be an *automorphism* of the fundamental group of S representing the mapping class ψ. Then for each n, the

K-quasi-isometry $\phi : M \to M$ from Step 1 has a unique extended lift

$$\Phi_n : (\mathbb{H}^3 \cup S_\infty^2) \to (\mathbb{H}^3 \cup S_\infty^2)$$

which is compatible with Ψ, in the sense that

$$\Phi_n \circ \rho_n(g) \circ \Phi_n^{-1} = \rho_n(\Psi(g))$$

for each $g \in \pi_1(S)$.

Pick $g_i \in \pi_1(S)$, $i = 1, 2, 3$, such that no pair commute. Then Φ_n maps the triple of attracting or parabolic fixed points for $(\rho_n(g_i))$ to the corresponding triple for $(\rho_n(\Psi(g_i)))$. As $n \to \infty$, these triples converge to triples of distinct points for ρ_∞. (The condition of noncommutativity assures the points are distinct). It follows that $\Phi_n | S_\infty^2$ is a precompact sequence of quasiconformal mappings, and that $\Phi_n | \mathbb{H}^3$ is a precompact sequence of K-quasi-isometries.

Any limiting map Φ_∞ satisfies

$$\Phi_\infty \circ \rho_\infty(g) \circ \Phi_\infty^{-1} = \rho_\infty(\Psi(g))$$

by continuity, and normalizes the geometric limit Π. Thus Φ_∞ descends to a quasi-isometry ϕ_∞ on M_∞ and ξ on N compatible with the covering projection π.

Step 5. Since the limit set of N is the whole sphere, and the injectivity radius of N is bounded above, Theorem 2.9 (Bounded rigidity) implies $\xi : N \to N$ is homotopic to an isometry. The map ϕ_∞ covers ξ, so it is also homotopic to an isometry $\alpha : M_\infty \to M_\infty$.

Step 6. The isometry group of a nonelementary hyperbolic manifold is discrete, and α has infinite order, so it acts freely and properly discontinuously on M_∞. The quotient $M_\infty / \langle \alpha \rangle$ has the homotopy type of T_ψ because α and ψ act the same way on the fundamental group. By a theorem of Stallings, the two manifolds are actually homeomorphic [St]. ∎

3.5 Asymptotic geometry

In the construction of the hyperbolic 3-manifold T_ψ fibering over the circle, we did not address the convergence of the two iterative processes involved. It was enough to obtain a convergent subsequence, by compactness of a Bers slice in one instance and by Thurston's double limit theorem in the other.

In this section we show both processes actually converge. When S is closed, we also show the totally degenerate group Γ on the boundary of a Bers slice enjoys a remarkable rigidity property: any quasiconformal conjugacy from Γ to another Kleinian group is differentiable at uncountably many points of the limit set. Finally we characterize the stable manifold of the fixed point of ψ.

Let $M_\psi \in AH(S)$ denote the covering space of the oriented hyperbolic manifold T_ψ corresponding to the inclusion of S into T_ψ as a fiber.

Proposition 3.10 *The mapping class ψ has exactly two fixed points in $AH(S)$, namely M_ψ and $M_{\psi^{-1}}$.*

Remark. These two fixed points are related by complex conjugation in $PSL_2(\mathbb{C})$.

Proof. If $\psi(P) = P$, then P is an infinite cyclic covering space of a finite volume hyperbolic manifold T homotopy equivalent to T_ψ. By Mostow rigidity, T is isometric to T_ψ. If the isometry preserves orientation, then $P = M_\psi$; otherwise $P = M_{\psi^{-1}}$. Finally M_ψ and $M_{\psi^{-1}}$ are distinct points in $AH(S)$; otherwise the mapping cylinders for ψ and ψ^{-1} would be identical, which would imply $\psi^2 = 1$ in $\mathrm{Mod}(S)$.

■

Theorem 3.11 (Convergence of iteration on $AH(S)$) *For any pseudo-Anosov mapping $\psi \in \mathrm{Mod}(S)$, and any $(X, Y) \in \mathrm{Teich}(S) \times \mathrm{Teich}(\overline{S})$, there exists a point $M_{\psi,Y} \in \partial B_Y$ such that*

$$Q(\psi^{-n}(X), Y) \rightarrow M_{\psi,Y} \quad and$$
$$\psi^n(M_{\psi,Y}) \rightarrow M_\psi$$

in $AH(S)$. The first limit point is independent of X, and the second is independent of Y.

In the course of the proof we will also establish:

Theorem 3.12 *The sequence $\psi^n(M_{\psi,Y})$ converges algebraically and geometrically to M_ψ. The negative end of the pared submanifold of $M_{\psi,Y}$ is asymptotically isometric to that of M_ψ.*

Here the *pared submanifold* of a hyperbolic manifold M is the complement of the open components of the thin part corresponding to cusps. By the tameness Theorem 3.5, the pared submanifold is homeomorphic to $S \times \mathbb{R}$ for any $M \in AH(S)$ without accidental parabolics. This homeomorphism is unique up to isotopy if we require it respect the orientations of M and $S \times \mathbb{R}$. Thus we may label the ends of the pared submanifold of M *positive* and *negative*, corresponding to the ends of $\mathbb{R}$.

The Theorem asserts the existence of a diffeomorphism

$$h : E_\psi \to E_{\psi,Y}$$

between the negative ends of the pared submanifolds of M_ψ and $M_{\psi,Y}$. The map h will be constructed in the natural homotopy class determined by the markings. The condition of *asymptotic isometry* means for any k, r and $\epsilon > 0$, there is a compact set $K \subset E_\psi$ such that h is ϵ-close to an isometry in the C^k topology on any embedded r-ball in $E_\psi - K$.

Note that when S is closed, there are no cusps and E_ψ is simply an end of M_ψ.

Corollary 3.13 *The manifold $M_{\psi,Y}$ does not have arbitrarily short closed geodesics.*

Proof. Any sequence of geodesics whose lengths tend to zero must tend to the negative end of $M_{\psi,Y}$. But the geometry of this end is asymptotically periodic, since it is quasi-isometric to an end of M_ψ.

∎

Corollary 3.14 *When S is closed, the injectivity radius of $M_{\psi,Y}$ is bounded above and below in its convex core.*

Proof. By Proposition 3.7, $M_{\psi,Y}$ has no cusps, and it has no short geodesics by the preceding Corollary, so the injectivity radius is bounded below. The upper bound (true for general surface groups) is Theorem 3.4.

∎

Corollary 3.15 *Suppose S is closed, and $M_{\psi,Y} = \mathbb{H}^3/\Gamma$. Then the limit set Λ of the totally degenerate group Γ contains an uncountable set of deep limit points D. Any quasiconformal conjugacy from Γ to another Kleinian group is differentiable and $C^{1+\alpha}$-conformal at each point of D.*

Proof. Let $\pi : M_\psi \to \mathbb{R}$ be the pullback of the smooth fibration $T_\psi \to S^1$, and let $\Sigma = \pi^{-1}(0)$. Consider any geodesic ray $\gamma(s)$ in M_ψ such that $\pi(\gamma(s)) \leq -\epsilon s < 0$ for some $\epsilon > 0$ and all large s. Then clearly $d(\gamma(s), \Sigma)$ tends to infinity at a linear rate with s. Thus the image δ of γ under an asymptotic isometry from M_ψ to $M_{\psi,Y}$ also penetrates the convex core of $M_{\psi,Y}$ at a linear rate, and therefore $\widetilde{\delta} \subset \mathbb{H}^3$ terminates at a deep limit point of Γ. Since the injectivity radius of $M_{\psi,Y}$ is bounded above and below in its convex core, any quasiconformal conjugacy is $C^{1+\alpha}$-conformal at every deep limit point, by Theorem 2.18. (The constant α depends on the limit point and the conjugacy.)

To produce such geodesic rays $\gamma(s)$, one can begin with rays which cover loops in T_ψ that have an essential projection to S^1. This gives countably many deep points. To obtain uncountably many, choose elements α and β generating a free group in $\pi_1(T_\psi)$, and mapping to negative elements in $\pi_1(S^1)$. The geodesic representative of any infinite positive word in higher powers of α and β produces a deep limit point, and words with different tails give different points. The set of possible tails is uncountable, so the set of deep limits points is also uncountable.

∎

Remark. These deep limit points are analogous to the points in the postcritical set $P(f)$ of an infinitely renormalizable quadratic polynomial; compare Theorems 8.8 and 8.13.

For reference we also record:

Theorem 3.16 (Cannon-Thurston) *The limit set of $M_{\psi,Y}$ is locally connected.*

See [CaTh] or [Min]. Corollary 8.3 below states an analogous result for renormalization.

Proof of Theorems 3.11 and 3.12. For simplicity we will initially assume that S is closed. Continuing with the notation of the previous section, let $M_0 = (f_0 : S \to M)$ denote an accumulation point of $Q(\psi^{-i}(X), Y)$ as $i \to \infty$, let M_∞ be an accumulation point of $M_n = \psi^n(M_0)$ as $n \to \infty$, and let (N, ω) be a geometric limit of (M_n, ω_n) covered by $(M_\infty, \omega_\infty)$. We now verify:

1. *The sequence M_n converges to M_∞.* The only possible accumulation points of M_n are M_ψ and $M_{\psi^{-1}}$. But the quasi-isometric distance from M_n to M_{n+1} is bounded, while the quasi-isometric distance from M_ψ to $M_{\psi^{-1}}$ is infinite, so the sequence can only accumulate at one of these two points.

2. *The algebraic and geometric limits $(M_\infty, \omega_\infty)$ and (N, ω) agree.* We have seen there are isometries α and β such that T_ψ is homeomorphic to $M' = M_\infty/\langle\alpha\rangle$ and the diagram

$$
\begin{array}{ccc}
M_\infty & \xrightarrow{\ \alpha\ } & M_\infty \\
\pi \downarrow & & \pi \downarrow \\
N & \xrightarrow{\ \beta\ } & N
\end{array}
$$

commutes. Since M' has finite volume, the covering map $M' \to N' = N/\langle\beta\rangle$ has finite degree d. But the covering $\pi : M_\infty \to N$ is just the pullback of the covering $M' \to N'$, so it also has degree d. For any $\delta \in \Pi = \pi_1(N)$, we have

$$\delta^d = \rho_\infty(\zeta) \in \Gamma_\infty = \pi_1(M_\infty),$$

and also $\delta = \lim \rho_n(\gamma_n)$ for a subsequence of n, where γ_n and ζ lie in $\pi_1(S)$. Then $\gamma_n^d = \zeta$ for n sufficiently large; since a surface group is uniquely divisible, γ_n eventually stabilizes in $\pi_1(S)$ and therefore $\delta \in \Gamma_\infty$.[2]

3. *The convex core of M is asymptotically isometric to M_∞.* Choose the marking $f_\infty : S \to \Sigma \subset M_\infty$ so its image is an embedded

---

[2]Cf. [Th1, Thm. 9.14].

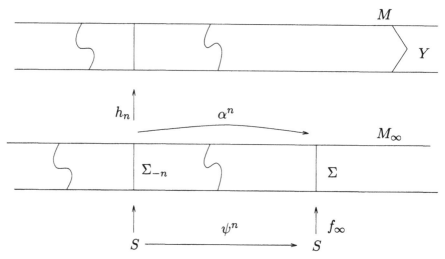

Figure 3.1. Construction of an asymptotic isometry.

surface corresponding to a fiber of $M_\infty/\langle\alpha\rangle$. We let $\Sigma_i = \alpha^i(\Sigma)$, and denote by $[\Sigma_i, \Sigma_j]$ the compact submanifold of M_∞ bounded by these two translates of the fiber.

Let $K = [\Sigma_{-3}, \Sigma_3]$; then $\Sigma \subset K$. By geometric convergence, for all n sufficiently large, there is a smooth embedding $g_n : K \to M$ such that $g_n \circ f_\infty$ is homotopic to $f_n = f_0 \circ \psi^{-n}$, and g_n converges towards an isometry in the C^∞ topology as $n \to \infty$.

Let
$$h_n = g_n \circ \alpha^n : [\Sigma_{-n-3}, \Sigma_{-n+3}] \to M.$$

The map h_n lies in the natural homotopy class of maps from M_∞ to M determined by the markings f_0 and f_∞ (see Figure 3.1). Indeed, $\alpha \circ f_\infty$ is homotopic to $f_\infty \circ \psi$, so if ρ is a retraction of M_∞ to the domain of h_n, then

$$h_n \circ \rho \circ f_\infty \sim h_n \circ \alpha^{-n} \circ f_\infty \circ \psi^n = g_n \circ f_\infty \circ \psi^n \sim f_0,$$

where $\sim$ denotes homotopy equivalence of mappings.

We will construct an asymptotic isometry h from one end of M_∞ into M by combining these h_n for large n. Let $\sigma : M_\infty \to [0, 1]$ be a smooth bump function supported on $[\Sigma_{-1}, \Sigma_1]$, such that

$\sum_{-\infty}^{\infty} \sigma(\alpha^k(x)) = 1$ for all x. For all x in $\bigcup_{n \gg 0}[\Sigma_{-n-1}, \Sigma_{-n}]$, let

$$h(x) = \sum_k \sigma(\alpha^k(x))h_k(x)$$

where the sum is interpreted as the hyperbolic barycenter of the weighted points. (The barycenter makes sense because all the maps are in the same homotopy class.)

To check that h is an asymptotic isometry, note that for any fixed x at most two terms, say $h_n(x)$ and $h_{n+1}(x)$, have nonzero weight. Both maps provide almost isometric embeddings of a definite neighborhood of Σ_{-n} into M, in the same homotopy class. Thus for all sufficiently large n, the images of h_n and h_{n+1} have substantial overlap, and the map

$$k_n = h_{n+1}^{-1} \circ h_n \mid [\Sigma_{-n-1}, \Sigma_{-n+1}]$$

is well-defined and homotopic to the identity. Therefore

$$\alpha^n \circ k_n \circ \alpha^{-n} : [\Sigma_{-1}, \Sigma_1] \to M_\infty$$

gives a sequence of almost isometric mappings homotopic to the identity. The only isometry of M_∞ homotopic to the identity *is* the identity, so $\alpha^n \circ k_n \circ \alpha^{-n} \to \mathrm{id}$ in the C^∞ topology. It follows that h_n and h_{n+1} are C^∞ close near x, and therefore a convex combination of these two maps is still C^∞ close to an isometry. This shows h is an asymptotic isometry from one end of M_∞ to the end of the convex core of $M_{\psi,Y}$.

Since $h(\Sigma_{-n})$ exits the geometrically infinite end of $M_{\psi,Y}$ (rather than the end bounded by Y), we have:

4. $M_\infty = M_\psi$.

To complete the proof we show:

5. For all $X \in \mathrm{Teich}(S)$, the sequence $Q(\psi^{-n}(X), Y)$ converges, and the limit is independent of X.

Let M_0 be an accumulation point of this sequence, and M_0' another accumulation point using X' in place of X. Since the geometrically infinite ends of M_0 and M_0' are both quasi-isometric to the same end of M_ψ, the corresponding groups Γ_0 and Γ_0' are quasiconformally conjugate. The geometrically finite ends of both manifolds correspond to the Riemann surface Y, so the conjugacy can be made

conformal on the domain of discontinuity. Since the injectivity radius of the convex core of M_0 is bounded above, the conjugacy is
conformal on the limit set; and therefore $M_0 = M_0'$, completing the
proof.

When $\partial S \neq \emptyset$, steps 1, 2 and 4 remain the same; in step 3
the asymptotic isometry is constructed in the pared submanifold of
M_∞; and in step 5 the mapping is extended over the cusps in a
straightforward way to show any two limits are quasi-isometric. ∎

A similar argument characterizes the attractor of M_ψ in $AH(S)$.

Theorem 3.17 (Stable manifold of M_ψ) *Let P belong to $AH(S)$.
Then $\psi^n(P) \to M_\psi$ if and only if the negative end of the pared submanifold of P admits an asymptotic isometry to M_ψ compatible with
markings.*

Proof. Let $h : E_\psi \to E$ be an asymptotic isometry between negative ends of M_ψ and P compatible with markings, and let $f : S \to
E_\psi \subset M_\psi$ be a marking of M_ψ. Then $h \circ \alpha^{-n} \circ f : S \to P$ is a
marking of $\psi^n(P)$. When n is large, the map h is nearly isometric on
$\alpha^{-n}(f(S))$, so $\psi^n(P)$ is algebraically close to $\psi^n(M_\psi) = M_\psi$. Thus
$\psi^n(P)$ converges to M_ψ.

Conversely, if $\psi^n(P) \to M_\psi$, a result of Thurston guarantees the
convergence is also geometric (see [Th1, Ch. 9], [Can, Thm. 9.2]).
Thus an asymptotic isometry between the ends of M_ψ and P can
be constructed by the partition of unity argument of the preceding
proof. ∎

3.6   Speed of algebraic convergence

Let S be a *closed* oriented surface of genus $g \geq 2$. In this section we apply geometric inflexibility to show, for any pseudo-Anosov
$\psi \in \mathrm{Mod}(S)$, the iterates $\psi^n(M_{Y,\psi})$ converge exponentially fast to

the fixed point M_ψ of ψ. This result, together with the characterization of the stable manifold of the fixed point above, forms the beginning of a dynamical theory of the action of $\mathrm{Mod}(S)$ on $AH(S)$.

Rapid algebraic convergence. To measure the speed of convergence, it is useful to regard points in $AH(S)$ as conjugacy classes of representations $\rho : \pi_1(S) \to \mathrm{Isom}^+(\mathbb{H}^3)$. We say $[\rho_n] \to [\rho]$ *exponentially fast* in $AH(S)$ if representations can be chosen within each equivalence class so that for each $\gamma \in \pi_1(S)$, there is a $\lambda < 1$ such that

$$\sup_{x \in S^2_\infty} |\rho_n(\gamma)(x) - \rho(\gamma)(x)| = O(\lambda^n).$$

(It suffices to check this condition on the generators of $\pi_1(S)$.)

A second, equivalent definition of exponential convergence can be based on traces: $[\rho_n] \to [\rho]$ exponentially fast if and only if for each $\gamma \in \pi_1(S)$ there is a $\lambda < 1$ such that

$$|\mathrm{tr}(\rho_n(\gamma))^2 - \mathrm{tr}(\rho(\gamma))^2| = O(\lambda^n).$$

Note that the complex function $\mathrm{tr}(\rho(\gamma))$ is defined only up to sign, since we are working in $PSL_2(\mathbb{C})$ instead of $SL_2(\mathbb{C})$. Again, it suffices to check this condition on finitely many elements in $\pi_1(S)$.

To relate the two definitions, let $R(S)$ denote the space of all *irreducible* representations $\rho : \pi_1(S) \to \mathrm{Isom}^+(\mathbb{H}^3)$. The space $AH(S)$ is a closed subset of the *representation variety*

$$V(S) = R(S)/\mathrm{Isom}^+(\mathbb{H}^3).$$

The space $V(S)$ is a complex manifold, which can be embedded in $\mathbb{C}^N$ using the holomorphic functions

$$[\rho] \mapsto \mathrm{tr}(\rho(\gamma))^2$$

for sufficiently many $\gamma \in \pi_1(S)$.

On a smooth manifold, there is a natural notion of exponential convergence using any local Euclidean chart. The first definition says $[\rho_n] \in V(S)$ admits a lift to a sequence ρ_n converging exponentially fast in the manifold $R(S)$. The second says $[\rho_n]$ converges exponentially fast with respect to smooth coordinates on $V(S)$. The two definitions are the same because the map $R(S) \to V(S)$ is a smooth fibration [Gun1, Thm. 28].

Theorem 3.18 (Rapid algebraic convergence) *For any pseudo-Anosov mapping $\psi \in \mathrm{Mod}(S)$, and any $Y \in \mathrm{Teich}(\overline{S})$, $\psi^n(M_{\psi,Y})$ converges to M_ψ exponentially fast.*

Corollary 3.19 *The derivative of the mapping class ψ has at least one expanding and one contracting eigenvalue at the fixed point M_ψ.*

Proof. The action of ψ on $V(\pi_1(S))$ is smooth, so exponential convergence entails an eigenvalue of modulus less than one at the fixed point. There is a symplectic structure on $V(\pi_1(S))$ preserved by $\mathrm{Mod}(S)$ [Gol], so we also have an expanding eigenvalue (since the determinant is 1). ∎

■

Question. Is the derivative of ψ at M_ψ actually hyperbolic (that is, free from eigenvalues of modulus 1)?

To give the proof of Theorem 3.18, we first show a deformation with a small quasi-isometry constant has a small effect on the traces. It is easy to bound the distortion of length, but we also need to know the twist along a closed geodesic is nearly preserved. (The combination of twist and length gives the trace.)

Lemma 3.20 *Let $M_i = \mathbb{H}^3/\langle\gamma_i\rangle$, $i = 1,2$, be a pair of hyperbolic solid tori, with oriented core geodesics g_i of length at most R. Suppose $\Phi : \mathcal{N}(g_1, 1) \to M_2$ is a $(1 + \epsilon)$-quasi-isometric embedding of a unit neighborhood of g_1 into M_2, such that $\Phi(g_1)$ is homotopic to g_2. Then*

$$|\operatorname{tr}(\gamma_1)^2 - \operatorname{tr}(\gamma_2)^2| \leq C(R)\epsilon.$$

Proof. Lift to the universal cover so γ_1 stabilizes a geodesic $\widetilde{g}_1$ through the origin 0 in the ball model for $\mathbb{H}^3$. Let $\widetilde{\Phi} : \mathcal{N}(\widetilde{g}_1, 1) \to \mathbb{H}^3$ be a corresponding lift of Φ, conjugating γ_1 to γ_2. Let K be the closure of $\mathcal{N}(\widetilde{g}_1, 1/2) \cap B(0, R+1)$. Since K is compact, by Theorem B.24 there is an isometry ι such that $d(\iota \circ \widetilde{\Phi}(x), x)C(R)\epsilon$ for all $x \in K$.

In other words, for an appropriate lift of γ_2, we can assume γ_1 and γ_2 are conjugate by a mapping $\widetilde{\Phi}$ which is ϵ-close to the identity on K.

Let v_i, $i = 1, \ldots, 4$, be the vertices of a regular tetrahedron of side length $1/4$ near 0. Then

$$d(\gamma_1(v_i), \gamma_2(v_i)) = O(\epsilon)$$

for all i, by virtue of the bound on $\tilde{\Phi}$. The map $\gamma \mapsto (\gamma(v_i))$ gives a smooth proper embedding of the isometry group of $\mathbb{H}^3$ into $(\mathbb{H}^3)^4$; by the implicit function theorem, γ can be smoothly reconstructed from the data $(\gamma(v_i))$, and therefore γ_1 and γ_2 are ϵ-close in $\mathrm{Isom}^+(\mathbb{H}^3)$. In particular the difference of their traces is $O(\epsilon)$.

■

Proof of Theorem 3.18 (Rapid algebraic convergence). We have seen that the homotopy class of ψ is realized by a quasi-isometry $\Psi : M_{\psi,Y} \to M_{\psi,Y}$. By Corollary 3.14, the injectivity radius of $M_{\psi,Y}$ in its convex core K is bounded above and below. By Theorem 2.11 (Geometric inflexibility), there are constants C and $\alpha > 0$, and a quasi-isometry

$$\Phi : M_{\psi,Y} \to M_{\psi,Y}$$

homotopic to Ψ, such that the quasi-isometry constant of Φ satisfies

$$L(\Phi, p) \;\le\; 1 + C \exp(-\alpha\, d(p, \partial K)) \tag{3.1}$$

for all $p \in K$. To show $\psi^n(M_{\psi,Y}) \to M_\psi$ exponentially fast, we will construct markings that move out the end of the convex core at a definite rate, and then apply the bound above.

Let $\alpha : M_\psi \to M_\psi$ be the isometry in the homotopy class of ψ. Let $f : S \to \Sigma$ be a representative marking of M_ψ, where Σ is an embedded surface disjoint from its translates $\Sigma_i = \alpha^i(\Sigma)$, $i \neq 0$. Then Σ represents a fiber of the surface bundle $M_\psi/\langle\alpha\rangle = T_\psi \to S^1$. As in the proof of Theorem 3.12, we let $[\Sigma_i, \Sigma_j]$ denote the compact submanifold bounded by two translates of the fiber; and we let $(-\infty, \Sigma_j] = \bigcup_{i<j}[\Sigma_i, \Sigma_j]$.

By Theorem 3.12, there is an asymptotic isometry $h : E \to E'$ between corresponding geometrically infinite ends of M_ψ and $M_{\psi,Y}$. We can assume that $E = (-\infty, \Sigma]$, and that E' is contained in the convex core K of $M_{\psi,Y}$.

Let $\Sigma'_i = h(\Sigma_i)$. In the periodic manifold M_ψ, the minimal distance between adjacent fibers is constant: $d(\Sigma_i, \Sigma_{i+1}) = d(\Sigma_0, \Sigma_1) =$

$D > 0$. Since h is an asymptotic isometry, we have $d(\Sigma_i', \Sigma_{i+1}') \geq D' > 0$ for all $i \leq 0$. Any path from Σ_i' to ∂K must cross Σ_j' for all j with $i < j \leq 0$, so we have:

$$d((-\infty, \Sigma_i'], \partial K) \;\geq\; D'|i|$$

for all $i < 0$.

Let $f_n : S \to M_{\psi,Y}$ be the marking given by the composition

$$S \xrightarrow{f} \Sigma \xrightarrow{\alpha^{-n}} \Sigma_{-n} \xrightarrow{h} \Sigma_{-n}' \subset M_{\psi,Y}.$$

Then $\psi^n(M_{\psi,Y}) = (f_n, M_{\psi,Y})$ as a sequence in $AH(S)$. Let $[\rho_n]$ be the corresponding sequence of representations of $\pi_1(S)$.

Picking any nontrivial element $\gamma \in \pi_1(S)$, we will now show $\mathrm{tr}(\rho_n(\gamma))^2 \to \mathrm{tr}\,\rho(\gamma))^2$ exponentially fast. Let g and g_n denote the geodesic representatives of γ in (f, M_ψ) and $(f_n, M_{\psi,Y})$. We claim there is a constant C such that $d(g_n, \partial K) \geq Cn > 0$ for all n sufficiently large. To see this, first note that g_n and $h(\alpha^{-n}(g))$ are homotopic, by the definition of f_n. Since h is an asymptotic isometry, these loops are very close when n is large. Now $g \subset (-\infty, \Sigma_N])$ for some N, so $h(\alpha^{-n}(g)) \subset (-\infty, \Sigma_{N-n}']$. But the distance from $(-\infty, \Sigma_{N-n}']$ to the boundary of the convex core grows linearly in n for n sufficiently large, so we have $d(g_n, \partial K) \geq Cn > 0$ as well.

Since Φ is in the homotopy class of ψ, the geodesic $\Phi(g_n)$ is homotopic to g_{n-1}. Combining our estimate for $d(g_n, \partial K)$ with equation (3.1), we find that on a unit neighborhood of g_n, the mapping Φ is a $(1 + \epsilon_n)$-quasi-isometry, where

$$\epsilon_n = O(\exp(-\alpha(d(g_n, \partial K)))) = O(\lambda^n)$$

for all $n \gg 0$, and $\lambda = \exp(-C\alpha) < 1$. By Lemma 3.20, this implies

$$|\,\mathrm{tr}(\rho_n(\gamma))^2 - \mathrm{tr}(\rho_{n-1}(\gamma))^2| \;=\; O(\lambda^n),$$

and thus $[\rho_n] \to [\rho]$ exponentially fast.

■

A canonical asymptotic isometry. By a similar argument, one can show that

$$f(x) = \lim_{n \to \infty} \Phi^n \circ h \circ \alpha^{-n}(x)$$

exists and defines a smooth map of M_ψ to $M_{\psi,Y}$. The mapping f satisfies $f \circ \alpha = \Phi \circ f$, and its quasi-isometry constant tends to one exponentially fast: that is, there is a $\lambda < 1$ such that $L(f,x) = 1 + O(\lambda^n)$ for all $x \in [\Sigma_{-n}, \Sigma_{-n+1}]$ and $n > 0$. The map f is independent of the choice of the initial asymptotic isometry h. It can be made canonical — that is, to depend only on ψ and Y — by using the Teichmüller mapping from Y to itself in the homotopy class of ψ to construct Φ.

Note. A detailed account of the representation variety of a surface group can be found in [Gun1], [Gol] and [Mor].

3.7 Example: torus bundles

The simplest surface with an interesting mapping class group is the torus. However, a torus bundle over the circle is never hyperbolic (its natural geometry is Euclidean, nilpotent or solvable). To obtain hyperbolic examples, we introduce a single cone point of angle $2\pi/p$, making the torus into a hyperbolic *orbifold* S. Then any $\psi \in \text{Mod}(S) = SL_2(\mathbb{Z})$ with $|\text{tr}(\psi)| > 2$ determines a hyperbolic 3-orbifold T_ψ which fibers over the circle.

In this section we study the corresponding totally degenerate groups $\Gamma_{\psi,Y}$ on the boundary of a Bers slice. We show the limit sets of these groups are deep at each elliptic fixed point. Any quasiconformal map $\phi : \widehat{\mathbb{C}} \to \widehat{\mathbb{C}}$ inducing the automorphism ψ of $\Gamma_{\psi,Y}$ provides a natural asymptotic self-similarity of the limit set, with a universal scaling factor. Analogous statements for the Julia set of an infinitely renormalizable quadratic map will appear in §9.4.

The problem of giving an explicit example of a totally degenerate group was raised in [Bers1]. This section also provides a numerical example of such a group, explains how it was calculated, and presents a picture of the limit set.

The orbifold bundles we consider here were studied by Jørgensen in [Jor]. Punctured torus groups, which can be thought of as the limiting case $p = \infty$, are discussed in [MMW] and [Wr]. Another example of a totally degenerate group, computed by a different method, appears in [MMW, §6]. For background on orbifolds, see [Th4, Ch. 5].

Hyperbolic torus bundles. Let S be the closed orbifold obtained from a surface of genus 1 by introducing a singular point of index $p > 1$. The group $G = \pi_1(S)$ has the presentation $\langle a, b : [a, b]^p = 1 \rangle$. We identify $H_1(S, \mathbb{Z}) = G/[G, G]$ with $\mathbb{Z}^2$ using the basis $\langle a, b \rangle$. Then the mapping class group $\mathrm{Mod}(S)$ is isomorphic to $SL_2(\mathbb{Z})$, faithfully represented by its action on $H_1(S)$, just as for a torus. Up to inner automorphism, the generators $R = \left(\begin{smallmatrix} 1 & 1 \\ 0 & 1 \end{smallmatrix}\right)$ and $L = \left(\begin{smallmatrix} 1 & 0 \\ 1 & 1 \end{smallmatrix}\right)$ of $\mathrm{Mod}(S)$ act on $\pi_1(S)$ by

$$R : \langle a, b \rangle \rightarrow \langle a, ba \rangle \quad \text{and}$$
$$L : \langle a, b \rangle \rightarrow \langle ab, b \rangle.$$

Every complex orbifold X in $\mathrm{Teich}(S)$ carries a hyperbolic metric and can be constructed by gluing together opposites sides of a hyperbolic quadrilateral with total internal angle $2\pi/p$. See Figure 3.2 for a tiling of the hyperbolic plane corresponding to the case $p = 2$.

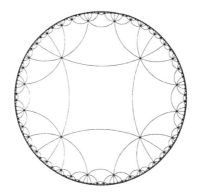

Figure 3.2. The universal cover of a torus with a 180° cone point.

The orbifold S admits a finite covering space $\widetilde{S} \rightarrow S$ which is a smooth surface. Explicitly, we can adjoin generators and relations to G to obtain the mod p Heisenberg group (of order p^3),

$$H = \langle a, b, c : c = [a, b], [a, c] = [b, c] = a^p = b^p = c^p = 1 \rangle.$$

Let $\tilde{S}$ be the covering space corresponding to the kernel of the natural map $G \to H$. Since $[a, b]$ has order p in H, $\tilde{S}$ is a manifold. For example, when $p = 2$, $\tilde{S}$ is a surface of genus three. Moreover, every $\psi \in \mathrm{Mod}(S)$ lifts to a map $\tilde{\psi} \in \mathrm{Mod}(\tilde{S})$; i.e. the covering is *characteristic.*

For $\psi \in \mathrm{Mod}(S)$, let T_ψ denote the three-dimensional orbifold fibering over the circle with fiber S and monodromy ψ. The singular locus of T_ψ is topologically a circle (see Figure 3.3).

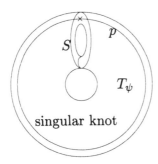

Figure 3.3. A fibered orbifold.

We claim *the orbifold T_ψ is hyperbolic if and only if ψ is pseudo-Anosov.* To see this, pass to the characteristic cover $\tilde{S}$ and apply Theorem 3.9 to obtain a hyperbolic structure on $T_{\tilde{\psi}}$. Then $T_{\tilde{\psi}}$ is a finite regular cover of T_ψ; by Mostow rigidity the covering group can be realized by hyperbolic isometries, so the hyperbolic structure descends to T_ψ.

The map ψ is pseudo-Anosov if and only if $|\mathrm{tr}(\psi)| > 2$ as an element of $SL_2(\mathbb{Z})$, so this property is easy to test.

Elliptic deep points. Next we examine a totally degenerate group $\Gamma_{\psi,Y}$, and show its limit set is asymptotically self-similar near any elliptic fixed point. We will apply results of the preceding sections which, while stated for surfaces, extend by naturality to the orbifold setting as well.

Let $M_{\psi,Y} = \mathbb{H}^3/\Gamma_{\psi,Y}$. Then $M_{\psi,Y}$ is a totally degenerate orbifold on the boundary of a Bers slice B_Y, $Y \in \mathrm{Teich}(\overline{S})$. The singular locus of $M_{\psi,Y}$ lies along a geodesic running from the orbifold point of Y into the geometrically infinite end of $M_{\psi,Y}$. Thus any primitive elliptic element $\delta \in \Gamma_{\psi,Y}$ has one fixed point in the limit set Λ and the other in the domain of discontinuity Ω. Normalize coordinates on the sphere so $\delta(z) = e^{2\pi i k/p}z$, and the limit set contains $z = 0$.

Let $\phi : Y \to Y$ be a quasiconformal map in the isotopy class of ψ. Then ϕ fixes the singular point of Y, and prolongs to a quasi-isometry Φ of $M_{\psi,Y}$, so there is a quasiconformal map $\tilde{\phi} : \widehat{\mathbb{C}} \to \widehat{\mathbb{C}}$ fixing 0 and ∞ and covering ϕ on Ω. The mapping ϕ is unique up to composition with a power of δ.

We can now verify:

1. *The elliptic fixed point $z = 0$ is a deep point of the limit set.* Indeed, the singular geodesic γ of M_ψ projects to a closed geodesic on T_ψ which runs once around, transverse to the fibers; thus $d(\gamma(s), \Sigma) \to \infty$ at a linear rate with respect to any reference fiber Σ in M_ψ. Since $M_{\psi,Y}$ is asymptotically isometric to M_ψ, its singular locus also tends to the geometrically infinite end at a linear rate. Thus any elliptic fixed point in the limit set of $\Gamma_{\psi,Y}$ is a deep point.

2. *The map $\tilde{\phi}$ is a quasiconformal automorphism of $\Gamma_{\psi,Y}$.* Indeed, $\tilde{\phi} \circ \gamma = \psi_*(\gamma) \circ \tilde{\phi}$. So by Theorem 2.18 we have:

3. *The map $\tilde{\phi}$ is $C^{1+\alpha}$-conformal at all deep points of the limit set.* In addition:

4. *We have $|\tilde{\phi}'(0)| > 1$.* To evaluate the derivative, recall the extension of ϕ to a map $\Phi : M_{\psi,Y} \to M_{\psi,Y}$ can be chosen to be an asymptotic isometry in the homotopy class of ψ. The negative end of $M_{\psi,Y}$ is nearly isometric to M_ψ, so Φ is asymptotic to the generator α of the deck group of the covering $M_\psi \to T_\psi$. This generator translates the singular locus towards the positive end of M_ψ, and therefore $|\tilde{\phi}'(0)| > 1$. Moreover:

5. *The self-similarity factor satisfies $|\tilde{\phi}'(0)| = e^L$, where L is the length of the closed singular geodesic on T_ψ.* Therefore:

6. *The self-similarity factor is universal, in the sense that it does not depend on the base surface Y.*

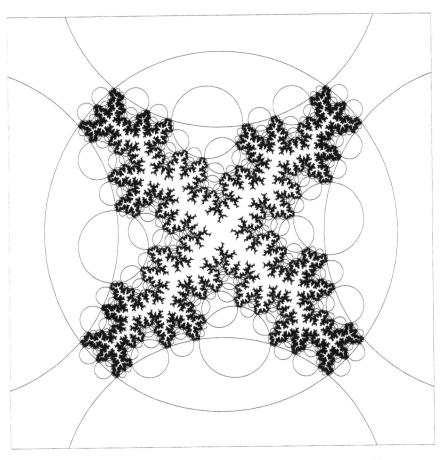

Figure 3.4. The limit set of a totally degenerate group $\Gamma_{\psi,Y}$.

At first sight, it may seem paradoxical that $\widetilde{\phi}(\Lambda) = \Lambda$ and $\widetilde{\phi}$ has an expanding fixed point at $z = 0$. Indeed, since $z = 0$ is a deep point, the limit set at a small scale looks much denser than at a large scale. The point is that $\widetilde{\phi}$ is not sufficiently smooth to expand small features of the limit set up to definite size. For example the iterates $\widetilde{\phi}^k$ are *not* uniformly quasiconformal near the origin.

A totally degenerate group. The assertions above are illustrated in an example in Figure 3.4. This Figure depicts a totally degenerate group corresponding to a representation

$$\rho : \pi_1(S) \to \Gamma_{\psi,Y}$$

with $p = 2$, $\psi = \left(\begin{smallmatrix} 2 & 1 \\ 1 & 1 \end{smallmatrix}\right)$ and $\operatorname{tr} \rho(a) = 3$. As in the preceding discussion, the group is normalized so $z = 0$ is an elliptic fixed point.

Because $\operatorname{tr} \rho(a)$ is real, the convex hull of $M_{\psi,Y}$ is pleated along the geodesic representing the simple curve C_a corresponding to a. The disk $D = \{z : |z| > 1\}$ is the boundary of a supporting half-space for the convex hull, and $\Omega = \bigcup_\Gamma \gamma(D)$. The Figure is drawn by enumerating the images of ∂D under Γ. These circles cut Ω into two types of tiles: lunes corresponding to the bending locus, and infinite-sided regions corresponding to the universal cover of $S - C_a$.

The origin is a limit of these circles, but since $z = 0$ is a deep point of the limit set, the circles become extremely small before they get close to the center of the Figure. Thus one should imagine the limit set as a nowhere dense but furry tree filling the central white region.

The four-fold symmetry of the picture is accounted for by the hyperelliptic involution $\langle a, b \rangle \to \langle a^{-1}, b^{-1} \rangle$.

To describe the calculation of this totally degenerate group, we first discuss how one can locate the fixed point M_ψ. It is convenient to work with representations in $SL_2(\mathbb{C})$ rather than $PSL_2(\mathbb{C})$. Let

$$\overline{G} = \langle a, b, c : c = [a, b]^p, c^2 = [a, c] = [b, c] = 1 \rangle$$

be the unique nontrivial $\mathbb{Z}/2$ central extension of $G = \pi_1(S)$. Any standard Fuchsian representation

$$\rho : G \to PSL_2(\mathbb{R})$$

lifts to a faithful representation

$$\overline{\rho} : \overline{G} \to SL_2(\mathbb{R}).$$

Although $\rho[a, b]$ is elliptic with rotation angle $2\pi/p$, the lifted representation has $\overline{\rho}([a, b]^p) = -I$ in $SL_2(\mathbb{R})$. Thus we have

$$\operatorname{tr} \overline{\rho}[a, b] = -\cos(2\pi/p). \tag{3.2}$$

Since the representations we are seeking are obtained by deforming the Fuchsian representations, they also enjoy these properties.

We claim the conjugacy classes of irreducible representations $\overline{\rho} : \overline{G} \to SL_2(\mathbb{C})$ satisfying (3.2) correspond bijectively to points on the affine variety

$$\alpha^2 + \beta^2 + \gamma^2 = \alpha\beta\gamma + 2 - 2\cos(2\pi/p) \tag{3.3}$$

in $\mathbb{C}^3$, via

$$(\alpha, \beta, \gamma) = (\operatorname{tr} \overline{\rho}(a), \operatorname{tr} \overline{\rho}(b), \operatorname{tr} \overline{\rho}(ab)).$$

To see this, first note that an irreducible representation $\overline{\rho}$ of the free group $\langle a, b \rangle$ is determined by the triple of traces above. Now in $SL_2(\mathbb{C})$ one has the identity

$$\operatorname{tr}(A)^2 + \operatorname{tr}(B)^2 + \operatorname{tr}(AB)^2 = \operatorname{tr}(A)\operatorname{tr}(B)\operatorname{tr}(AB) + 2 + \operatorname{tr}[A, B],$$

so points on the variety (3.3) correspond to representations with $\operatorname{tr} \overline{\rho}[a, b] = -2\cos(2\pi/p)$. But then $\overline{\rho}[a, b]$ is elliptic of order $2p$, so $\overline{\rho}$ gives a representation of $\overline{G}$ satisfying (3.2).

In terms of these coordinates, the generators of $\operatorname{Mod}(S)$ act by

$$\begin{aligned} R(\alpha, \beta, \gamma) &= (\alpha, \gamma, \alpha\gamma - \beta) \quad \text{and} \\ L(\alpha, \beta, \gamma) &= (\gamma, \beta, \beta\gamma - \alpha), \end{aligned}$$

as can be verified using the identity

$$\operatorname{tr} A^2 B + \operatorname{tr} B = \operatorname{tr} A \operatorname{tr} AB$$

(coming from $A + A^{-1} = (\operatorname{tr} A)I$).

It is now straightforward to solve for the fixed points of $\psi = R \circ L$ on the variety (3.3). The solution corresponding to M_ψ is $(\alpha, \beta, \gamma) = (\lambda, \bar{\lambda}, \lambda)$, where

$$\lambda = \frac{3 + \sqrt{17} - \sqrt{2 - 2\sqrt{17}}}{4};$$

its complex conjugate gives $M_{\psi^{-1}}$. The other two solutions are real, so they do not lie in $AH(S)$. The expanding eigenvalue of $D\psi$ at its fixed point M_ψ is

$$1 + \frac{\sqrt{17} + \sqrt{17 + 4\sqrt{17}}}{2} = 5.955184721953\ldots$$

A similar calculation can be carried out for other values of p and ψ; see [Jor].

To find the representation corresponding to the totally degenerate group $\Gamma_{\psi,Y}$, one may use Newton's method to solve numerically for the nearby intersection of the stable manifold of M_ψ with the hyperplane $\mathrm{tr}(\bar{\rho}(a)) = \alpha = 3$. When normalized as in Figure 3.4, the resulting group has generators

$$A = \begin{pmatrix} \frac{3+3i}{2} & -i\sqrt{\frac{7}{2}} \\ i\sqrt{\frac{7}{2}} & \frac{3-3i}{2} \end{pmatrix} \quad \text{and}$$

$$B \approx \begin{pmatrix} 3.9999415228 - 0.62754422329i & -3.3746078723 - 1.8636013648i \\ -2.8699108556 - 3.0395681614i & 0.62754422329 + 3.9999415228i \end{pmatrix}.$$

We remark that the orbifold T_ψ is obtained by Dehn filling on the figure eight knot complement, since the latter is a punctured torus bundle with monodromy $\left(\begin{smallmatrix} 2 & 1 \\ 1 & 1 \end{smallmatrix}\right)$. The singular geodesic of T_ψ has length $L \approx 0.7328576759$, and zero torsion; thus the totally degenerate limit set in Figure 3.4 is asymptotically self-similar about the origin with scale factor $e^L \approx 2.0810189966$.

Another totally degenerate representation of the same group, with $\mathrm{tr}(\bar{\rho}(a)) = \alpha = 3$ and $\psi = \left(\begin{smallmatrix} 5 & 4 \\ 1 & 1 \end{smallmatrix}\right)$, is rendered in Figure 3.5.

A proof that for each pseudo-Anosov ψ and $\alpha > 2$, there exists a unique totally degenerate group of the form $M_{\psi,Y}$ with convex hull bent along $\langle a \rangle$ and $\mathrm{tr}(\bar{\rho}(a)) = \alpha$, can be found in [Mc6].

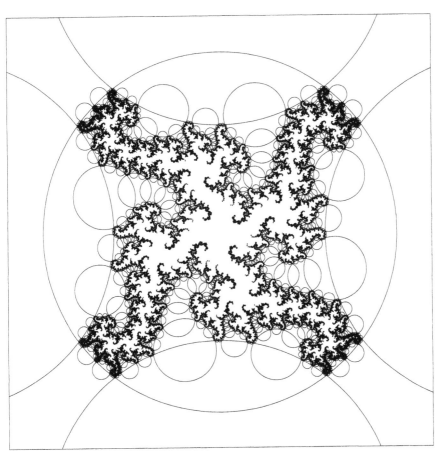

Figure 3.5. A totally degenerate group on the stable manifold of M_ψ, for $\psi = \left(\begin{smallmatrix} 5 & 4 \\ 1 & 1 \end{smallmatrix}\right)$.

4 Quadratic maps and renormalization

We now turn from Kleinian groups and mapping classes to quadratic maps and renormalization.

Let $f(z) = z^2 + c$ be a quadratic polynomial. The most interesting behavior of f is associated with the orbit of its critical point $z = 0$. Suppose an iterate f^n maps a disk U_n containing 0 over itself properly by degree two, with image V_n. Then $f^n : U_n \to V_n$ is a *quadratic-like map*, of the same general form as f itself. If the critical point remains in U_n under iteration of f^n, then $f^n : U_n \to V_n$ is a *renormalization* of f^n. The map f is *infinitely renormalizable* if there are arbitrarily high iterates that can be renormalized.

Our goal in the next several chapters is to understand the limiting form of $f^n : U_n \to V_n$ as n tends to infinity.

This chapter presents basic compactness and continuity properties of quadratic-like maps that are essential for working in this general setting. The picture of renormalization we describe is developed more fully in [Mc4].

4.1 Topologies on domains

Recall that the set of closed subsets of the complex plane $\mathbb{C}$ is compact in the Hausdorff topology (§2.2).

A *pointed region* (U, u) is an open connected set $U \subset \mathbb{C}$ equipped with a basepoint $u \in U$. The *Carathéodory topology* on the space of all pointed regions is defined by $(U_n, u_n) \to (U, u)$ if and only if

(i) $u_n \to u \in U$, and

(ii) for any Hausdorff limit F of a convergent subsequence of $\mathbb{C} - U_n$, U is a component of $\mathbb{C} - F$.

The Carathéodory topology is designed so that convergence of simply connected regions is the same as locally uniform convergence of their Riemann mappings [Car]. One may also verify:

Proposition 4.1 *With respect to the Carathéodory topology:*

1. *For any $u \in \mathbb{C}$ and $r > 0$, the set of pointed regions (U, u) such that $B(u, r) \subset U$ is compact.*

2. *The limit of a sequence of simply-connected regions is simply-connected.*

3. *If $(U_n, u_n) \to (U, u)$, and U_n and U are hyperbolic, then the hyperbolic metric ρ_n on U_n converges to the hyperbolic metric ρ on U. The convergence is C^∞ on compact sets.*

Consider next the space of all holomorphic functions $f : (U, u) \to \mathbb{C}$. The Carathéodory topology on this space is defined as follows: a sequence $f_n : (U_n, u_n) \to \mathbb{C}$ converges to $f : (U, u) \to \mathbb{C}$ if $(U_n, u_n) \to (U, u)$ and if $f_n \to f$ uniformly on compact subsets of U. (Note that any compact subset of U is also contained in U_n for all n sufficiently large.)

A *K-quasicircle* $\Lambda \subset \mathbb{C}$ is the image of a round circle under a K-quasiconformal map of the plane. Such a curve satisfies the *bounded turning condition*: for any $x, y \in \Lambda$,

$$\operatorname{diam} \Lambda_{xy} \leq M(K)|x - y|,$$

where Λ_{xy} is the smaller arc on Λ joining x to y. Conversely, a Jordan curve with bounded turning is a quasicircle (see [LV]).

A *disk* $U \subset \widehat{\mathbb{C}}$ is a simply-connected region with $U \neq \widehat{\mathbb{C}}$. We say U is a *K-quasidisk* if ∂U is a K-quasicircle.

Let (U_n, u_n) be a sequence of K-quasidisks in $\mathbb{C}$ converging to (U, u), where $U \neq \mathbb{C}$. Then U is also a K-quasidisk, and $\overline{U_n} \to \overline{U}$ in the Hausdorff topology. This last property does *not* hold for general disks, because islands can pinch off in the limit; the bounded turning condition prevents such pinching.

4.2 Polynomials and polynomial-like maps

For a polynomial $f : \mathbb{C} \to \mathbb{C}$ of degree $d > 1$, the *filled Julia set* $K(f)$ is the compact set of points which remain bounded under iteration; its boundary is the *Julia set* $J(f)$.

A *polynomial-like map* $f : U \to V$ is a proper holomorphic map between disks U and V in $\mathbb{C}$ such that $\overline{U}$ is compact and contained in V. By analogy with polynomials, the filled Julia set of f is defined by $K(f) = \bigcap f^{-n}(U)$; and the Julia set, by $J(f) = \partial K(f)$.

A polynomial-like map is a more flexible dynamical system retaining many of the features of a polynomial. To put both kinds of mappings together, we let $\mathcal{P}oly_d$ denote the space of all polynomial-like maps $f : (U, u) \to (V, v)$ and all polynomials $f : (\mathbb{C}, u) \to (\mathbb{C}, v)$ of degree d, with connected Julia sets and basepoints $u \in K(f)$. We give $\mathcal{P}oly_d$ the Carathéodory topology.

Remark. The notation $f : (U, u) \to (V, v)$ means $V = f(U)$ and $v = f(u)$. When $f_k : (U_k, u_k) \to (V_k, v_k)$ converges to $f : (U, u) \to (V, v)$ in $\mathcal{P}oly_d$, it can be shown that the pointed regions (V_k, v_k) tend to (V, v) (compare [Mc4, Thm. 5.6]).

For any $f \in \mathcal{P}oly_d$, the *postcritical set* $P(f)$ is the closure of the strict forward orbits of the critical points in U:

$$P(f) \;=\; \overline{\bigcup_{n \geq 1,\ f'(c)=0} f^n(c)}.$$

Note that we have adopted the convention that the postcritical set of a polynomial $f \in \mathcal{P}oly_d$ does *not* contain the point at infinity. This convention is consistent with the viewpoint that a polynomial is a limiting case of a polynomial-like map, and that the domain of a polynomial is the complex plane.

The following result is well-known for polynomials [Dou3], and easily generalizes to polynomial-like maps.

Proposition 4.2 *As a function on $\mathcal{P}oly_d$, the filled Julia set $K(f)$ varies upper semicontinuously, while $J(f)$ varies lower semicontinuously.*

In other words, if $f_k \to f$, then

$$\limsup K(f_k) \;\subset\; K(f) \;\;\text{and}$$
$$J(f) \;\subset\; \liminf J(f_k).$$

Thus if $K(f) = J(f)$, both sets vary continuously at f.

The *modulus* $\operatorname{mod}(E, V)$ of a set $E \subset V \subset \mathbb{C}$, where V is a simply-connected region, is the maximum modulus of an annulus $A \subset V$ surrounding E. (This means E lies in the compact component of $V - A$.) If no such annulus exists we set $\operatorname{mod}(E, V) = 0$.

The space $Poly_d(m) \subset Poly_d$ consists of all polynomials, and all polynomial-like maps $f : (U, u) \rightarrow (V, v)$ with $\mod(U, V) \geq m$.

The group $\mathrm{Aut}(\mathbb{C})$ of automorphisms $\alpha(z) = az + b$ of the complex plane acts continuously on $Poly_d$ by conjugation, sending $f(z)$ to $\alpha(f(\alpha^{-1}(z)))$. The subspace $Poly_d(m)$ is preserved by this action, and we have (cf. [Mc4, Theorem 5.8]):

Proposition 4.3 *The space $Poly_d(m)/\mathrm{Aut}(\mathbb{C})$ is compact. More precisely, any sequence $f_n : (U_n, u_n) \rightarrow (V_n, v_n)$ in $Poly_d(m)$, normalized so $u_n = 0$ and so the Euclidean diameter of $K(f_n)$ is equal to 1, has a convergent subsequence.*

4.3 The inner class

In this section we discuss conjugacies between quadratic-like maps. In particular, we define the *inner class* $I(f)$ of $f \in Poly_2$, a fundamental invariant introduced by Douady and Hubbard to record the dynamics on $K(f)$. The ideas in this section are based on [DH], to which the reader may refer for more details.

Definitions. Let $f_i : U_i \rightarrow V_i$, $i = 1, 2$ be a pair of polynomial-like maps of the same degree. A *hybrid conjugacy* is a quasiconformal map ϕ between neighborhoods of $K(f_1)$ and $K(f_2)$, conjugating f_1 to f_2, with $\bar{\partial}\phi = 0$ a.e. on $K(f_1)$. We say f_1 and f_2 are *hybrid equivalent* if such a conjugacy exists.

Proposition 4.4 *Any polynomial-like map f is hybrid equivalent to a polynomial $g : \mathbb{C} \rightarrow \mathbb{C}$ of the same degree as f. The polynomial g is unique up to affine conjugacy when $K(f)$ is connected.*

See [DH, Theorem 1].

Let $P_c(z) = z^2 + c$. The *Mandelbrot set* M is the set of $c \in \mathbb{C}$ such that $J(P_c)$ is connected.

A *quadratic-like map* is a polynomial-like map of degree two. Any quadratic-like map f with connected Julia set is hybrid equivalent to P_c for a unique $c \in M$.

The *inner class* $I(f)$ is defined by $I(f) = c$.

Let $f_i : U_i \rightarrow V_i$, $i = 1, 2$, be a pair of quadratic-like maps with connected Julia sets. The *pullback argument* is a construction for building a conjugacy ϕ between f_1 and f_2. It proceeds in stages, as follows.

1. After a slight restriction, we can assume U_i and V_i are disks bounded by smooth circles.

2. Pick an orientation-preserving diffeomorphism $\phi : \partial V_1 \to \partial V_2$.

3. Lift ϕ using the degree two covering maps $f_i : \partial U_i \to \partial V_i$, to obtain an extension to a conjugacy

$$\phi : (\partial U_1 \cup \partial V_1) \to (\partial U_2 \cup \partial V_2).$$

4. Extend ϕ to a diffeomorphism between the closed annuli $A_i = \overline{V}_i - U_i$, $i = 1, 2$, bounded by these circles. Since ϕ is smooth, its dilatation K is bounded.

5. Each annulus $V_i - K(f_i)$ is tiled by the annuli $A_i^k = f_i^{-k}(A_i)$, $k = 0, 1, 2 \ldots$ (see Figure 4.1). Define $\phi : A_1^k \to A_2^k$ inductively by $\phi = f_2^{-1} \circ \phi \circ f_1$, using the fact that $f_i : A_i^k \to A_i^{k-1}$ is a covering map. (The lift is chosen to agree with ϕ as already defined on the outer boundary of A_i^k.) Since f_i is conformal, the dilatation of ϕ remains the same during this step.

The final step, extending ϕ to a conjugacy on the filled Julia sets, can be completed if $I(f_1) = I(f_2)$:

Proposition 4.5 *A quasiconformal conjugacy $\phi : (V_1 - K(f_1)) \to (V_2 - K(f_2))$ extends across the filled Julia sets to a hybrid conjugacy if and only if f_1 and f_2 have the same inner class.*

Proof. Suppose f_1 and f_2 have the same inner class c. Using hybrid conjugacies from $f_i(z)$ to $z^2 + c$, we can reduce to the case where $f_1(z) = f_2(z) = z^2 + c$. Pulling back ϕ by the normalized Riemann mapping

$$\psi : (\mathbb{C} - \overline{\Delta}) \to (\mathbb{C} - K(f_1)),$$

we obtain a quasiconformal homeomorphism $\tilde{\phi}$ defined near S^1 and satisfying $\tilde{\phi}(z^2) = \tilde{\phi}(z)^2$. Any such mapping extends to the identity on the circle. This implies ϕ extends by the identity on $K(f_1)$ to

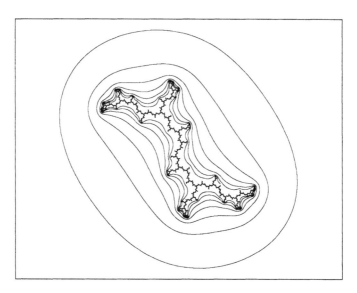

Figure 4.1. Fundamental annuli for a quadratic-like map.

a topological conjugacy. Extensions by the identity preserve quasi-conformality (cf. [DH, Lemma 2]), so we obtain a hybrid conjugacy extending ϕ.

The converse is immediate.

■

Proposition 4.6 *Let $f_i \in \mathcal{P}oly_2(m)$, $i = 1, 2$ be quadratic-like maps with the same inner class. Then for any $\epsilon > 0$, there are restrictions g_i of f_i in $\mathcal{P}oly_2(m - \epsilon)$, and a $K(m, \epsilon)$-quasiconformal map $\phi : \mathbb{C} \to \mathbb{C}$ providing a hybrid conjugacy between g_1 and g_2.*

Proof. Given f_1 and f_2, choose slight restrictions such that $A_i = \overline{V_i - U_i}$ is a smoothly bounded annulus. Pick a smooth map $\phi : A_1 \to A_2$ which is a conjugacy on the boundary, and prolong ϕ to a hybrid conjugacy using the pullback construction. Since ∂V_i is a quasicircle, ϕ extends to a quasiconformal homeomorphism of the whole plane.

The dilatation of ϕ is bounded by its dilatation on A_1. By continuity, we can control the modulus of A_i and the dilatation of ϕ for all pairs of maps on a neighborhood of (f_1, f_2) in $Poly_2(m) \times Poly_2(m)$. The Proposition follows by compactness of $Poly_2(m)/\operatorname{Aut}(\mathbb{C})$.

■

Proposition 4.7 *The inner class* $I : Poly_2 \to M$ *is continuous.*

Proof. The proof is a straightforward adaptation of [DH, Prop. 14].

Let $f : (U, u) \to (V, v)$ be a quadratic-like map in $Poly_2$. The inner class $I(f)$ can be constructed by a variant of the pullback argument.

Pick a disk $D \subset U$ with smooth boundary and compact closure in U, such that D contains $K(f)$ and $f : D_f \to D$ is quadratic-like, where $D_f = f^{-1}(D)$. Pick a diffeomorphism

$$\psi : (\mathbb{C} - D) \to \{z \; : \; |z| \geq 4\},$$

and extend it to a quasiconformal mapping

$$\psi_f : (\mathbb{C} - K(f)) \to (\mathbb{C} - \overline{\Delta})$$

such that $\psi_f(f(z)) = S(\psi(z))$ for $z \in D_f$, where $S(z) = z^2$. The map ψ_f is constructed by the pullback argument: that is, by lifting $\psi|\partial D$ to a map on ∂D_f, filling in with a smooth map from $\overline{D} - D_f$ to $\{z \; : \; 2 \leq |z| \leq 4\}$, and prolonging by the dynamics.

Using ψ_f we can construct a quadratic polynomial by gluing the dynamics of $S(z)$ onto the outside of D. More precisely, let μ_f denote the Beltrami differential which vanishes on $K(f)$ and is equal to the complex dilatation of ψ_f on the rest of $\mathbb{C}$. Let $\phi : \mathbb{C} \to \mathbb{C}$ be a quasiconformal mapping with dilatation μ_f, as provided by the measurable Riemann mapping theorem. Then

$$P(z) \;\; = \;\; \begin{cases} \phi \circ f \circ \phi^{-1}(z) & z \in \phi(D), \\ \phi \circ \psi_f^{-1} \circ S \circ \psi_f \circ \phi^{-1}(z) & \text{otherwise,} \end{cases}$$

is a quadratic polynomial, hybrid equivalent to f; so $P(z)$ is conformally conjugate to $z^2 + c$ where $c = I(f)$.

Now let W be a bounded open set with $\overline{D} \subset W \subset \overline{W} \subset U$, and let Λ be the space of holomorphic maps $g : W \to \mathbb{C}$ such that $|g(z) - f(z)| < \epsilon$ throughout W. Note that Λ is an infinite-dimensional complex manifold, in fact an open domain in a complex Banach space. Choosing ϵ sufficiently small, we may assume each g admits a quadratic-like restriction $g : D_g \to D$, where $D_g = g^{-1}(D)$. Let $\Lambda^{\text{conn}} \subset \Lambda$ denote the set of g such that $K(g|D_g)$ is connected. Any map h in $\mathcal{P}oly_2$ sufficiently close to f agrees with some $g \in \Lambda^{\text{conn}}$ on W. The straightening procedure just described can be carried out for all mappings g in Λ^{conn}.

Here are two basic features of the construction. Suppose $g_n \to g_\infty$ in the family Λ^{conn}, $I(g_n) = c_n$ and $I(g_\infty) = c_\infty$. Then:

(a) the polynomial $z^2 + c_\infty$ is quasiconformally conjugate to $z^2 + c$ for any accumulation point c of c_n; and

(b) if the indicator function of $K(g_n)$ converges to that of $K(g_\infty)$ in $L^1(\mathbb{C})$, then $c_n \to c_\infty$.

Feature (a) follows from compactness properties of quasiconformal maps. The hypothesis of (b) implies μ_{g_n} can be chosen to converge in measure to μ_{g_∞}, which implies the corresponding solutions to the Beltrami equation converge as well, and thus $I(g_n) \to I(g_\infty)$. These principles generalize to polynomials of any degree.

We will now show $g_n \to g_\infty$ implies $c_n \to c_\infty$, using a property special to degree two: a quadratic polynomial with an indifferent cycle is quasiconformally rigid.

Let $\mathcal{I} \subset \Lambda^{\text{conn}}$ denote the set of g with an indifferent periodic cycle, and let $\Lambda^{\text{stable}} = \Lambda - \overline{\mathcal{I}}$. Then $K(g)$ moves by a holomorphic motion over Λ^{stable} ([MSS], [Mc4, §4]). Thus the indicator function of $K(g)$ varies continuously in $L^1(\mathbb{C})$ (using absolute continuity of quasiconformal maps). So by (b) above, $c_n \to c_\infty$ when $g_\infty \notin \overline{\mathcal{I}}$.

Now suppose $g_n \to g_\infty \in \overline{\mathcal{I}}$. Then $g_\infty = \lim h_n$ where h_n has an indifferent cycle. Therefore $I(h_n)$ lies in the boundary of the Mandelbrot set, so the same is true of any accumulation point c' of $I(h_n)$. But then $z^2 + c'$ is quasiconformally rigid, so by (a) we have $c' = c_\infty$. Similarly (a) implies $c = c_\infty$ for any accumulation point c of c_n, completing the proof of continuity of $I : \Lambda^{\text{conn}} \to M$.

Since Λ^{conn} accounts for a neighborhood of f in $\mathcal{P}oly_2$, we have established continuity of the inner class on $\mathcal{P}oly_2$.

■

Remark. One can define an inner class on $\mathcal{P}oly_d$, but it is *not* continuous for $d \geq 3$ [DH].

4.4 Improving polynomial-like maps

As we have seen in the preceding section, it is often beneficial to tame a polynomial-like map by restricting it. This section presents a general construction which results in a restriction with several convenient properties.

The proof depends on basic facts that will also be used frequently in the sequel.

Proposition 4.8 (Modulus of a neighborhood) *Let $X \subset \mathbb{C}$ be a connected set of diameter 1; then:*

1. *For any $\epsilon > 0$, X is surrounded by an annulus of modulus at least $m(\epsilon) > 0$ contained in an ϵ-neighborhood of X.*

2. *Any open set $U \supset X$ with $\mathrm{mod}(U, X) \geq m > 0$ contains an $\epsilon(m)$-neighborhood of X, where $\epsilon(m) > 0$.*

Proof. (1) We may assume $\epsilon < 1$. Let N be the complement of X in an ϵ-neighborhood of X. The outer boundary $\gamma \subset N$ of an $\epsilon/2$-neighborhood of X has Euclidean length $O(1/\epsilon)$. The hyperbolic metric on N is comparable to $|dz|/d(z, \partial N)$, so the hyperbolic length of γ is $O(1/\epsilon^2)$. A standard collar neighborhood of the geodesic representative of γ in N provides the requisite annulus (cf. [Mc4, Thm. 2.19]).

(2) Suppose $x_1 \in X$ and $z \in \mathbb{C} - U$. Choose $x_2 \in X$ such that $|x_1 - x_2| \geq 1$. Then the hyperbolic length of a simple loop in $\widehat{\mathbb{C}} - \{x_1, x_2, z, \infty\}$ separating $\{x_1, x_2\}$ from $\{z, \infty\}$ tends to infinity as the cross-ratio $|z - x_1|/|x_1 - x_2|$ tends to zero; but it is bounded above in terms of $m = \mathrm{mod}(U, X)$. Thus $|z - x_1| \geq \epsilon(m) > 0$.

■

Here is a slight modification of [Mc4, Thm. 2.25]:

Proposition 4.9 (Inclusion contraction) *Let $\iota : X \hookrightarrow Y$ be an inclusion of one hyperbolic Riemann surface into another, and let $s = d(x, Y - X)$. Then with respect to the hyperbolic metrics on X and Y,*

$$\|\iota'(x)\| < C(s) < 1,$$

where $C(s)$ decreases to zero as $s \to 0$.

Proof. By passing to universal coverings and applying the Schwarz lemma, we can reduce to the extremal case $Y = \Delta$, $X = \Delta^* = \Delta - \{0\}$, $x > 0$ and $s = d(0, x)$ in the hyperbolic metric on the unit disk Δ. Thus we may take

$$C(s) = \frac{\rho_\Delta(x)}{\rho_{\Delta^*}(x)} = \frac{2|x \log x|}{1 - x^2} < 1,$$

where ρ_Δ and ρ_{Δ^*} denote the hyperbolic metrics on the disk and punctured disk. As s decreases to zero, so does x, and so does the right-hand expression above; thus $C(s)$ decreases to zero as well.

∎

Proposition 4.10 (Improving polynomial-like maps) *Let $f : U \to V$ be a polynomial-like map of degree d with a connected Julia set of diameter 1. Suppose V contains an ϵ-neighborhood of $K(f)$. Then there is a restriction $f : U' \to V'$ which is also polynomial-like of degree d, and such that:*

1. $\mathrm{mod}(U', V') \geq m > 0$;

2. *U' and V' are K-quasidisks; and*

3. $\mathrm{diam}(V') \leq C \, \mathrm{diam} \, K(f)$.

Here m, K and C depend only on (d, ϵ).

Remark. By Proposition 4.8, V contains an ϵ-neighborhood of $K(f)$ where $\epsilon > 0$ is determined by $\mathrm{mod}(U, V)$. Thus the same result holds where m, K and C depend only on d and $\mathrm{mod}(U, V)$.

Proof. We will give a concrete construction of U' and V'. In the course of the proof, all constants will depend only on ϵ and d.

It suffices to prove the Proposition when $\text{mod}(K(f), V)$ is bounded above. Indeed, there is a constant M such that $\text{mod}(K(f), V) \geq M$ implies V contains an ϵ-neighborhood of $K(f)$. So if $\text{mod}(K(f), V) \geq dM$, we can replace V with U and U with $f^{-1}(U)$, and the hypotheses will still be satisfied, because $\text{mod}(K(f), U) = (1/d)\,\text{mod}(K(f), V)$. After iterating this process a finite number of times we obtain $\text{mod}(K(f), V) < dM$.

Consider next the hyperbolic Riemann surfaces $U_0 \subset V_0$ obtained by doubling $U - K(f)$ and $V - K(f)$ across their ends corresponding to $K(f)$. Then U_0 and V_0 are annuli with the same core geodesic γ (which parameterizes the prime ends of $K(f)$), and $f|(U - K(f))$ extends to a symmetric mapping $F : U_0 \to V_0$ sending γ to itself (see Figure 4.2).

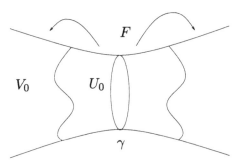

Figure 4.2. Double of a polynomial-like map.

Now impose the hyperbolic metric on V_0. Since $\text{mod}(V_0) = 2\,\text{mod}(K(f), V)$ is bounded above and below, the hyperbolic length of γ is also bounded above and below. Thus there is a constant R such that the collar of radius R about γ has modulus at least $(3/4)\,\text{mod}(V_0)$. Since $\text{mod}(U_0) = \text{mod}(V_0)/d \leq \text{mod}(V_0)/2$, we have $d(\gamma, V_0 - U_0) < R$. Thus $\|F'(z)\| > \lambda > 1$ for all points $z \in U_0$ within distance 1 of γ, and some constant λ. (Indeed, F is an isometry from the hyperbolic metric on U_0 to the hyperbolic metric on V_0, and the contraction of the inclusion $U_0 \hookrightarrow V_0$ at z is bounded in terms of the distance from z to $(V_0 - U_0)$ by the preceding Proposition.)

Let $V_0' \subset V_0$ be a unit collar neighborhood of γ, and let $U_0' = F^{-1}(V_0')$. Then U_0' is contained in a $1/\lambda$-neighborhood of γ, so the two annuli formed by $\{z \ : \ d(z, \gamma) \in [1, 1/\lambda]\}$ surround U_0' in V_0'. These annuli have definite moduli since the length of γ is bounded above. Thus $U' = (U_0' \cap U) \cup K(f)$ and $V' = (V_0' \cap V) \cup K(f)$ provide the domain and range of a polynomial-like restriction of f with definite modulus.

To check the quasidisk condition, we use the fact for any annulus $A \subset \mathbb{C}$, the core curve of A is a $K(m)$-quasicircle, where $m = \text{mod}(A)$. Since the length of γ is bounded above, there is an annulus $A \subset V_0 - \gamma$ of definite modulus invariant under the circular symmetry of V_0 and with $\partial V'$ as its core curve. Then $A \subset V \subset \mathbb{C}$, so we have established the quasidisk condition for V'; since $\partial U'$ is the core curve of $f^{-1}(A)$, U' is also a K-quasidisk with controlled K.

To obtain the diameter bound, first recall our preliminary normalization to ensure that that $\text{mod}(K(f), V)$ is bounded above. This implies the diameter of $K(f)$ is bounded below in the hyperbolic metric on V. On the other hand, $\text{mod}(V', V)$ is bounded below, since $A \subset V$, so $\text{diam} \, V' \leq C' \, \text{diam} \, K(f)$ in the hyperbolic metric on V. A bound of the same form for the Euclidean metric follows from the Koebe distortion theorem.

∎

4.5 Fixed points of quadratic maps

A quadratic polynomial has two fixed points, counted with multiplicity. These points can be labeled in a natural way when the Julia set is connected. This labeling extends to quadratic-like maps, as discussed in this section.

Let $f(z) = z^2 + c$ be a quadratic polynomial with connected Julia set. Let

$$\psi : (\mathbb{C} - \overline{\Delta}) \to (\mathbb{C} - K(f))$$

be the Riemann mapping, normalized so $\psi(z)/z \to 1$ at infinity.

An *external ray* R_t is the image under ψ of the ray $(1, \infty) \exp(2\pi i t)$. The external ray R_0 lands at a fixed point of f which we denote $\beta(f)$. The other fixed point of f is denoted $\alpha(f)$. When $c = 1/4$ we have

$\alpha(f) = \beta(f)$ and the fixed point is parabolic; otherwise $\alpha(f) \neq \beta(f)$ and $\beta(f)$ is repelling (cf. [Mc4, §6]).

Proposition 4.11 *Let $f(z) = z^2 + c$ be a quadratic polynomial with connected Julia set, and let p be a fixed point of f. Suppose there is a closed arc $\gamma \subset \mathbb{C}$ with one endpoint at p, such that $\gamma \subset f(\gamma)$ and $\gamma \cap K(f) = \{p\}$. Then $p = \beta(f)$.*
Conversely, there exists such an arc in any neighborhood of $\beta(f)$.

Proof. Consider $\psi^{-1}(\gamma)$; by Lindelöf's theorem, this arc terminates at a point $q \in S^1$, and since $f(\gamma) \supset \gamma$ we have $q^2 = q$. It follows that $q = 1$ and thus p is the landing point of the external ray R_0, so $p = \beta(f)$.
For the converse, let γ be the tail of R_0.

∎

For a quadratic-like map $f \in \mathcal{P}oly_2$, we define $\alpha(f)$ and $\beta(f)$ to be the fixed points of f corresponding to the α and β fixed points of $z^2 + c$, where $c = I(f)$. By the result above, $\beta(f)$ is topologically distinguished from $\alpha(f)$.

Proposition 4.12 *The fixed points $\alpha(f)$ and $\beta(f)$ are continuous functions of $f \in \mathcal{P}oly_2$.*

Proof. If $p = \beta(f)$ is parabolic then $\alpha(f) = \beta(f)$ and p is a fixed point of multiplicity two. Thus both fixed points of g lie close to p when g is close to f, establishing continuity.
Otherwise $\beta(f)$ is repelling, and by transversality any g near f has fixed points near $\alpha(f)$ and $\beta(f)$; we need only show that the fixed point near $p = \beta(f)$ is labeled $\beta(g)$.
There is a branch of f^{-1} fixing p and mapping a small ball B centered at p strictly inside itself. An invariant arc γ touching p from outside $K(f)$ can be described as the closure of a sequence of arcs

$$\delta \cup f^{-1}(\delta) \cup f^{-2}(\delta) \cup \cdots$$

matched end-to-end by f^{-1} and converging to p, where δ is a closed arc contained in B and disjoint from $K(f)$.

Since $K(f)$ varies upper-semicontinuously, δ is outside $K(g)$ for g close to f, and a suitable branch of g^{-1} still maps B strictly inside itself. A small modification δ' of δ (so the endpoints still match) permits one to construct an invariant arc

$$\delta' \cup g^{-1}(\delta') \cup g^{-2}(\delta') \cup \dots$$

tending to a fixed point q of g near p. Thus $q = \beta(g)$ by the arc characterization of the β fixed point.

∎

Here is another viewpoint on the proof above. Given a repelling fixed point p for f, transversality gives a nearby repelling fixed point p_g for all g in a neighborhood U of f in $\mathcal{P}oly_2$. Form a bundle of complex tori

$$T_g = (B - \{p_g\})/(z \sim g(z))$$

over U, where B is a small ball about p; this is the quotient space for the repelling dynamics near p. Since $K(g)$ is g-invariant, it descends to a compact subset $K_g \subset T_g$.

There is a distinguished isotopy class of simple closed curve ξ_g on T_g which lifts to a simple closed curve around p_g. We can now reformulate our characterization of the β fixed point as follows: $p_g = \beta(g)$ if and only if there is a simple closed curve $\gamma_g \subset T_g - K_g$ having intersection number 1 with ξ_g.

If $p_f = \beta(f)$, we can find such a curve γ_f, and then transport it by local triviality to a curve γ_g in the nearby fibers. For g near f, it remains disjoint from K_g by upper semicontinuity of the filled Julia set, and its intersection number with ξ_g remains 1. Thus $p_g = \beta(g)$ for all g near f, establishing continuity of $\beta(g)$.

4.6   Renormalization

Let $f : (U, u) \to (V, v)$ belong to $\mathcal{P}oly_2$, and let c be the critical point of f. We say f^n is *renormalizable* if there exist open disks $U_n \subset V_n \Subset \mathbb{C}$ such that $c \in U_n$ and

$$f^n : U_n \to V_n$$

is a quadratic-like map with connected Julia set. Then $f^n : (U_n, c) \rightarrow (V_n, f(c))$ also belongs to $\mathcal{P}oly_2$.

The choice of U_n and V_n is a *renormalization* of f^n. The Julia set and postcritical set of $f^n : U_n \rightarrow V_n$ will be denoted $J_n(f)$ and $P_n(f)$ respectively; they do not depend on the particular choice of U_n and V_n.[1]

The *small Julia sets* at level n are defined by

$$J_n(i) = f^i(J_n(f)), \quad i = 1, 2, \ldots, n.$$

Let $V_n(i) = f^i(U_n)$ and let $U_n(i)$ be the component of $f^{i-n}(U_n)$ contained in $V_n(i)$; then

$$f^n : U_n(i) \rightarrow V_n(i), \quad i = 1, 2, \ldots, n$$

are quadratic-like maps with Julia sets $J_n(i)$. As i varies, the maps so defined are all conformally conjugate.

The *small filled Julia sets* K_n and $K_n(i)$ are defined similarly. The *small postcritical sets* are defined by $P_n(i) = P(f) \cap K_n(i)$; their union is $P(f)$.

When necessary we will denote the dependence of these sets on f by $J_n(f)$, $J_n(i, f)$ etc.

Example. The first frame of Figure 4.3 depicts the filled Julia set of $f(z) = z^2 + c$ where $c \approx -1.13000 + 0.24033i$; the critical point of this mapping has period six. The second frame illustrates the domain, range and filled Julia set of a renormalization $f^2 : U_2 \rightarrow V_2$. The renormalized map is hybrid equivalent to "Douady's rabbit", with inner class $c \approx -0.22561 + 0.744862i$.

The next result shows that the small Julia sets are not too small.

Proposition 4.13 *Let f^n be renormalizable, where $f : U \rightarrow V$ is in $\mathcal{P}oly_2(m)$. Then for $1 \leq i \leq n$, we have*

$$1 \geq \frac{\text{diam } J_n(i)}{\text{diam } J(f)} \geq C(m, n) > 0.$$

---

[1]It is also common to refer to $f^n : U_n \rightarrow V_n$, if it exists, as the nth renormalization of f.

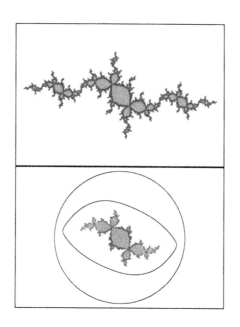

Figure 4.3. Renormalization.

Proof. First suppose $f(z) = z^2 + c$. Let $B(0,r)$ be the smallest ball containing J_n. Then $f(J_n) \subset B(c, r^2)$. Since $f^i(J_n) \subset J(f) \subset B(0,2)$, and $|f'(z)| = 2|z|$, we can make the estimate

$$2r = \operatorname{diam} J_n = \operatorname{diam} f^n(J_n) \leq 2 \cdot 10^n r^2;$$

thus $r \geq 10^{-n}$ and $\operatorname{diam} J_n / \operatorname{diam} J(f) > 10^{-n}$. A similar bound holds for $J_n(i)$.

To treat the case of polynomial-like maps $f \in \mathcal{Poly}_2(m)$, apply Proposition 4.6 and the distortion theorems for quasiconformal mappings. ∎

4.7 Simple renormalization

We now recall the notion of simple renormalization. If two distinct small Julia sets $J_n(i)$ and $J_n(j)$ meet, they do so at a repelling fixed point p of f^n. This fixed point can be classified as type α or β depending on whether it is the α or β fixed point of $f^n : U_n(i) \to V_n(i)$. It is known that all intersections are of the same type. Accordingly, we classify the renormalization of f^n as:

α-type, if the small Julia sets meet at their α-fixed points;

β-type, if they meet at their β-fixed points; and

disjoint type, if the small Julia sets are disjoint.

Definition. A renormalization is *simple* if it is of disjoint type or β-type. Equivalently, simple means that $K_n(i) - K_n(j)$ is connected for any $i \neq j$ (cf. [Mc4, §7]).

We let

$$\mathcal{R}(f) = \{n \geq 1 : f^n \text{ is renormalizable}\}, \text{ and}$$
$$\mathcal{SR}(f) = \{n \geq 1 : f^n \text{ is simply renormalizable}\}.$$

All renormalizations of real quadratic polynomials are simple, as are the usual renormalizations constructed with the Yoccoz puzzle (mapping a single puzzle piece over itself). If f^a and f^b are both simply renormalizable, and $a < b$, then a divides b, $J_b \subset J_a$ and for a suitable choice of U_b, $f^b|U_b$ is a renormalization of $f^a|U_a$. (These properties are *not* true in general without the assumption of simplicity.)

We will be concerned exclusively with simple renormalizations in the sequel.

4.8 Infinite renormalization

A mapping $f : (U, u) \rightarrow (V, v)$ in $Poly_2$ is *infinitely renormalizable* if $\mathcal{SR}(f)$ is infinite. (This is equivalent to the condition that $\mathcal{R}(f)$ is infinite [Mc4, Theorem 8.4].)

Let $\text{mod}(f, n)$ be the supremum of $\text{mod}(U_n, V_n)$ over all simple renormalizations of f^n. We set $\text{mod}(f, n) = 0$ if f^n is not renormalizable, and $\text{mod}(f, 1) = \infty$ if f is a polynomial.

For $n > 1$ such that f^n is renormalizable, the quantity $\text{mod}(f, n)$ is positive and finite, and there exists a renormalization $f^n : U_n \rightarrow V_n$ with $\text{mod}(U_n, V_n) = \text{mod}(f, n)$.

Proposition 4.14 *Let f be infinitely renormalizable, and suppose*

$$\limsup_{n \in \mathcal{SR}(f)} \text{mod}(f, n) > 0.$$

Then the diameters of the small Julia sets of f tend to zero; that is, $\sup_{1 \leq i \leq n} \text{diam } J_n(i) \rightarrow 0$ as $n \rightarrow \infty$ in $\mathcal{SR}(f)$.

Proof. By assumption, there is an $m_0 > 0$ such that $\text{mod}(f, n) \geq m_0$ for infinitely many n. By Proposition 4.10 (and the Remark following), for each such n we can choose a renormalization such that $\text{mod}(U_n, V_n) \geq m > 0$, $\text{diam } V_n \leq C \text{ diam } J_n(f)$, and U_n and V_n are K-quasidisks. Here m, C and K depend only on m_0.

Thus we have $V_n \subset B(c, R)$, where c is the critical point of f and $R = C \text{ diam } J(f)$ is independent of n. Consequently $U_n \subset f^{-n}(B(c, R))$. But the sets $f^{-n}(B(c, R))$ nest down to the filled Julia set $K(f)$, and since f is infinitely renormalizable, the filled Julia set is nowhere dense (that is, $K(f) = J(f)$). Thus the radius of the largest round ball contained in U_n tends to zero as n tends to infinity. But the diameter of a K-quasidisk is bounded in terms of the diameter of the largest ball it contains, so diam $U_n \to 0$.

This shows diam $J_n \to 0$ as $n \to \infty$ along the subsequence where $\mathrm{mod}(f, n) \geq m_0$. Similar reasoning shows $\sup_i \mathrm{diam}\, J_n(i)$ also tends to zero along this subsequence. But $\sup_i J_n(i)$ is a decreasing function of $n \in \mathcal{SR}(f)$ (since the Julia sets at high levels nest inside those at lower levels), so $\sup_i J_n(i) \to 0$ along the full sequence of $n \in \mathcal{SR}(f)$.

■

Under the same hypotheses, one may further establish that $J(f)$ is locally connected at the critical point of f. Compare [JH], [Ji].

Proposition 4.15 *Suppose $f_k \to f$ in $\mathcal{P}oly_2$, and for each k, f_k^n is simply renormalizable and $\mathrm{mod}(f_k, n) \geq m > 0$. Then:*

1. *f^n is simply renormalizable;*

2. *$\mathrm{mod}(f, n) \geq m$;*

3. *$\limsup_k K_n(f_k) \subset K_n(f)$; and*

4. *$J_n(f) \subset \liminf_k J_n(f_k)$.*

Proof. We have diam $J(f) = \lim \mathrm{diam}\, J(f_k)$, so diam $J(f_k)$ is bounded above and below; thus diam $J_n(f_k)$ is also bounded above and below, by Proposition 4.13.

Let c_k and c denote the critical points of f_k and f; then $c_k \to c$. Choose renormalizations

$$f_k^n : U_n^k \to V_n^k$$

with $f_k^n : (U_n^k, c_k) \to \mathbb{C}$ in $\mathcal{P}oly_2(m)$. By Proposition 4.3, we can pass to a convergent subsequence, and obtain in the limit a renormalization $f^n : (U_n, c) \to \mathbb{C}$ in $\mathcal{P}oly_2(m)$.

Recall that the Julia set $J_n(f)$ is independent of the choice of renormalization. Thus $J(f^n|U_n)$ is the same for any limit of the sequence $f_k^n : (U_n^k, c_k) \to \mathbb{C}$. So from lower semicontinuity of $J(\cdot)$ as a function on $\mathcal{P}oly_2$, we obtain $J_n(f) \subset \liminf J_n(f_k)$. A similar argument gives upper semicontinuity of $K_n(f_k)$.

Finally we check that the renormalization of f^n is simple. If not, then two small Julia sets $J_n(i, f)$ and $J_n(j, f)$ meet at single point p, a repelling fixed point of f^n which is also the α fixed point of both small Julia sets. By continuity of $\alpha(f)$ on $\mathcal{P}oly_2$, the α fixed points of $J_n(i, f_k)$ and $J_n(j, f_k)$ both converge to p as $k \to \infty$. But since p is repelling, it is a fixed point of multiplicity one, and so f_k^n has only one fixed point near p for all k sufficiently large. Thus the α fixed points of $J_n(i, f_k)$ and $J_n(j, f_k)$ coincide for large k, and these renormalizations are not simple either.

∎

5 Towers

We now broaden our scope and consider *towers* of quadratic-like maps compatible under renormalization. These towers arise naturally as geometric limits of infinitely renormalizable quadratic polynomials. (This statement is made precise in §9.4.)

5.1 Definition and basic properties

A *level set* $S \subset \mathbb{Q}^+$ is a collection of positive rational numbers containing 1, such that for any pair $s < t$ in S, t/s is an integer.

A *tower* $\mathcal{T}$ is a collection of quadratic-like maps

$$\langle f_s : U_s \to V_s; \ s \in S \rangle$$

such that:

1. S is a level set;

2. the critical point of each $f_s(z)$ is at $z = 0$;

3. the filled Julia set $K(f_s)$ is connected; and

4. for any levels s, t with $n = t/s > 1$, the map f_s^n is simply renormalizable, $K(f_t) = K_n(f_s)$, and $f_s^n = f_t$ on $K(f_t)$.

Two examples.
A. Let $f \in \mathcal{P}oly_2$ be infinitely renormalizable. Let $S = \mathcal{SR}(f)$, and let $f_s = f^s : U_s \to V_s$ be a renormalization of f^s. Then $\mathcal{T} = \langle f_s; \ s \in S \rangle$ is a tower.

B. Suppose $f : U \to V$ is a quadratic-like fixed point of renormalization, meaning there is a $p > 1$ and an $\alpha \in \mathbb{C}^*$ such that

$$f(z) = \alpha f^p(\alpha^{-1} z)$$

for all $z \in K(f)$. Let $S = \{p^n : n \in \mathbb{Z}\}$, and let $f_{p^n} = \alpha^{-n} f \alpha^n$. Then $\mathcal{T} = \langle f_s; \ s \in S \rangle$ is also a tower.

In example (A), $V_s \subset V_1$ for all s. However, even in this example one does not generally have $V_t \subset V_s$ for all $t > s$; we have not required

95

any nesting condition in our definition of the renormalizations of a quadratic polynomial. Condition (4) in the definition of a tower is formulated to allow for this greater flexibility; we do *not* required that $V_t \subset V_s$ when $t > s$, even when $s = 1$. In example (B) there may be no relation between V_s and V_t.

The following Proposition allows nesting to be recovered after a controlled loss of modulus.

Proposition 5.1 *Given levels $s < t$, let U and V be the components of $U_s \cap U_t$ and $V_s \cap V_t$ containing $K(f_t)$. Then $f_t = f_s^{t/s}$ on U and the restriction $f_t : U \to V$ is a simple renormalization of $f_s^{t/s}$.*

There is a further restriction $f_t : U' \to V'$ which also satisfies $\mathrm{mod}(U', V') \geq m' > 0$, where m' depends only on a lower bound for $\mathrm{mod}(U_s, V_s)$ and $\mathrm{mod}(U_t, V_t)$.

Proof. Since $f_t = f_s^{t/s}$ on $K(f_t)$, this equality holds throughout U. The restriction $f_t : U \to V$ is proper because f_t and $f_s^{t/s}$ are proper maps. Since $K(f_t) \subset U$, the restricted map is still quadratic-like, and its filled Julia set is unchanged. By the definition of towers, $K(f_t) = K_{t/s}(f_s)$, so the restriction is a simple renormalization of $f_s^{t/s}$.

A lower bound m on $\mathrm{mod}(U_s, V_s)$ and $\mathrm{mod}(U_t, V_t)$ guarantees that both V_s and V_t contain an $\epsilon(m) \cdot \mathrm{diam}(K(f_t))$ neighborhood of $K(f_t)$, so the same is true of V. Then by Proposition 4.10, there is a further restriction $f_t : U' \to V'$ with $\mathrm{mod}(U', V') \geq m'(\epsilon(m)) > 0$. ∎

We topologize the space *Tow* of all towers as follows. First, a sequence of level sets S_k converges to S if the indicator function of $S_k \subset \mathbb{Q}^+$ converges pointwise to S. In other words:

1. S contains all s which belong to S_k for infinitely many k, and

2. every $s \in S$ belongs to all but finitely many S_k.

Then a sequence

$$\mathcal{T}_k = \langle f_{s,k} : U_{s,k} \to V_{s,k}; \ s \in S_k \rangle$$

converges to $\mathcal{T} = \langle f_s : U_s \to V_s; \ s \in S \rangle$ if in addition:

3. for each $s \in S$ and for all k sufficiently large, $f_{s,k}$: $(U_{s,k}, 0) \to \mathbb{C}$ converges to $f_s : (U_s, 0) \to \mathbb{C}$ in the Carathéodory topology.

Conjugating all the maps in a tower by the change of coordinates $z \mapsto \alpha z$ determines another tower; thus we have an action of $\mathbb{C}^*$ on $\mathcal{T}ow$. (Note that the full group $\mathrm{Aut}(\mathbb{C})$ does not act, because the critical point of each f_s must remain at $z = 0$.)

Definite moduli. A tower has *definite moduli* if $\inf \mathrm{mod}(U_s, V_s) > 0$. Let $\mathcal{T}ow(m)$ denote the space of all towers such that $\mathrm{mod}(U_s, V_s) \geq m$ for all $s \in S$.

Proposition 5.2 *The space $\mathcal{T}ow(m)/\mathbb{C}^*$ is compact. More precisely, any sequence $\mathcal{T}_k$ in $\mathcal{T}ow(m)$ normalized so that $\mathrm{diam}\, J(f_{1,k}) = 1$ has a convergent subsequence.*

Proof. It is easy to see the space of level sets is compact, so after passing to a subsequence we can arrange that $S_k \to S$. By assumption, for each $s \in S$ the maps $f_{s,k}$ (defined for k sufficiently large) all lie in $\mathcal{P}oly_2(m)$. Since $\mathcal{P}oly_2(m)/\mathrm{Aut}(\mathbb{C})$ is compact, and the diameter of $J(f_{1,k})$ is normalized to be 1, we obtain convergence of $f_{1,k}$ after passing to a further subsequence. But by Proposition 4.13, the diameter of $J(f_{1,k})$ controls the diameter of $J(f_{s,k})$, so after passing to a further subsequence we may assume $f_{s,k}$ converges to a map $f_s : (U_s, 0) \to \mathbb{C}$ in $\mathcal{P}oly_2(m)$, for all $s \in S$.

Now consider levels t and s in S with $n = t/s > 1$. By Proposition 5.1, $\mathrm{mod}(f_{s,k}, n) \geq m' > 0$ for all k. Thus f_s^n is also renormalizable, and the remaining properties required for a tower follow from Proposition 4.15.

∎

5.2 Infinitely renormalizable towers

The space $\mathcal{L}ev$ of all level sets S forms a Cantor set in the topology introduced above, and the natural map $\mathcal{T}ow \to \mathcal{L}ev$ is continuous. When t is a level of $\mathcal{T} = \langle f_s; s \in S \rangle$, we define $\mathcal{T}$ *shifted by t* as the tower $\mathcal{T}' = \langle g_s; s \in S' \rangle$, where $S' = \{s/t : s \in S\}$ and $g_s = f_{ts}$.

After shifting by t, f_t becomes g_1. Shifting by t is continuous on the closed and open subset of $\mathcal{T}ow$ where it is defined.

A tower $\mathcal{T} = \langle f_s; \ s \in S \rangle$ is *renormalizable* if it has a level s with $s > 1$. Its *renormalization* is $\mathcal{T}$ shifted by the least level $s > 1$. We say $\mathcal{T}$ is *infinitely renormalizable* if this process can be repeated indefinitely; equivalently, if $\sup S = \infty$. Clearly if $\mathcal{T}$ is infinitely renormalizable then f_s is infinitely renormalizable for every $s \in S$.

Let $\mathcal{T}ow^\infty(m)$ denote the space of all towers with $\sup S = \infty$ and $\mathrm{mod}(U_s, V_s) \geq m$ for all $s \in S$.

Proposition 5.3 *Both $J(f_1)$ and $P(f_1)$ are continuous functions on $\mathcal{T}ow^\infty(m)$.*

Proof. Since f_1 is infinitely renormalizable, $J(f_1) = K(f_1)$. Then continuity of $J(f_1)$ follows from upper and lower semicontinuity of $K(f)$ and $J(f)$ for $f \in \mathcal{P}oly_2$.

Now suppose $\mathcal{T}_k \to \mathcal{T}$ in $\mathcal{T}ow^\infty(m)$; passing to a subsequence, we can assume $P(f_{1,k}) \to Q$ in the Hausdorff topology, for some compact set Q. It is easy to see that $Q \supset P(f_1)$. To complete the proof we will show $Q \subset P(f_1)$.

By Proposition 5.1, $\mathrm{mod}(f_1, s) \geq m'(m) > 0$ for all $s \in S$ with $s > 1$. Thus f_1 is infinitely renormalizable, with definite moduli at infinitely many levels. By Proposition 4.14, the diameters of the small Julia sets $J_s(i, f_1)$ tend to zero as $s \to \infty$. Given $\epsilon > 0$, choose $s \in S$ such that $\mathrm{diam}\, J_s(i, f_1) < \epsilon$ for $i = 1, \ldots, s$.

The small Julia sets at level s vary continuously, and cover the postcritical set; this implies Q is contained in a small neighborhood of $P(f_1)$. More precisely, for all k large enough, s belongs to S_k and

$$P(f_{1,k}) \subset J_s(1, f_{1,k}) \cup \ldots \cup J_s(s, f_{1,k}).$$

We have $J_s(i, f_{1,k}) \to J_s(i, f_1)$ for each i, so for k large enough, $P(f_{1,k})$ is within an ϵ-neighborhood of the union of the small Julia sets at level s for f_1. But each $J_s(i, f_1)$ contains a point in the postcritical set $P(f_1)$, and has diameter less than ϵ, so the union of the small Julia sets at level s for f_1 lies within an ϵ-neighborhood of $P(f_1)$. Thus Q is within 2ϵ of $P(f_1)$; since ϵ was arbitrary, $Q \subset P(f_1)$ and we are done.

■

Corollary 5.4 *If s is in the level set of a tower $T \in Tow^\infty(m)$, then $J(f_s)$ and $P(f_s)$ are defined and continuous on a neighborhood of T in $Tow^\infty(m)$.*

Proof. That s is in the level set of nearby towers follows from the definition of the topology on Tow. The Corollary follows from the preceding result by shifting the level set so f_s becomes f_1.

∎

Remark. The postcritical set $P(f_c)$ is only lower semicontinuous on the space of all quadratic polynomials. For example, the postcritical set of $f(z) = z^2 - 2$ consists of the 2 points $\{-2, 2\}$, but there are polynomials $f_k(z) = z^2 - c_k$, $c_k \to 2$, such that $P(f_k)$ is the full interval $[-c_k, c_k]$.

5.3 Bounded combinatorics

Two distinct levels s and t of a tower are *adjacent* if there is no level between s and t. A tower has *bounded combinatorics*, bounded by B, if $t/s \leq B$ for all adjacent levels $t > s$. Let $Tow^\infty(m, B) \subset Tow^\infty(m)$ denote the space of all infinitely renormalizable towers with $\mathrm{mod}(U_s, V_s) \geq m$ and with combinatorics bounded by B.

Theorem 5.5 *The space $Tow^\infty(m, B)/\mathbb{C}^*$ is compact.*

Proof. This is immediate from compactness of $Tow(m)/\mathbb{C}^*$ and compactness of the levels sets with $\sup S = \infty$ and combinatorics bounded by B.

∎

Note that $Tow^\infty(m)/\mathbb{C}^*$ itself is *not* compact; for example, a sequence of infinite level sets S_k can converge to the trivial level set $S = \{1\}$.

The combination of compactness and continuity leads to bounds.

Proposition 5.6 *Let T be a tower in $Tow^\infty(m, B)$, and let n be the least level of T greater than 1. Then there is definite space between* the Julia set $J(f_n)$ and the part of the postcritical set $P(f_1)$ it does not contain; that is,

$$\frac{d(J(f_n), P(f_1) - P(f_n))}{\operatorname{diam} J(f_n)} \geq \epsilon(m, B) > 0.$$

Proof. Since f_1 is infinitely renormalizable, the postcritical set $P(f_1)$ can be written as the disjoint union

$$P(f_1) = P(f_n) \cup f_1(P(f_n)) \cup \ldots f_1^{n-1}(P(f_n)),$$

where $P(f_n) \subset J(f_n)$ and the remaining sets are disjoint from $J(f_n)$ [Mc4, Thm. 8.1]. Thus the ratio above is positive for any given tower $\mathcal{T}$. The numerator can be expressed as the minimum of $d(J(f_n), f_1^i(P(f_n)))$ over $i = 1, \ldots, n$; since $J(f_n)$ and $P(f_n)$ vary continuously in the Hausdorff topology, the numerator varies continuously at $\mathcal{T}$. The same is true of the denominator, so the ratio is bounded away from zero on a neighborhood of $\mathcal{T}$. Compactness of $Tow^\infty(m, B)/\mathbb{C}^*$ then implies a lower bound valid for all $\mathcal{T}$.

■

We can immediately generalize this bound:

Proposition 5.7 *For any pair of levels $s < t$ for $\mathcal{T}$ in $Tow^\infty(m, B)$, we have*

$$\frac{d(J(f_t), P(f_s) - P(f_t))}{\operatorname{diam} J(f_t)} \geq \epsilon(m, B) > 0.$$

Proof. Write the levels between s and t as $s = s_0 < s_1 < \ldots s_n = t$. By the preceding result and the shift invariance of $Tow^\infty(m, B)$, there is an $\epsilon(m, B) > 0$ such that $W_i \cap P(f_{s_{i-1}}) = P(f_{s_i})$, where W_i is an $\epsilon(m, B) \cdot \operatorname{diam}(J(f_{s_i}))$-neighborhood of $J(f_{s_i})$. Clearly $W_1 \supset W_2 \supset \ldots W_n$, and therefore $W_n \cap P(f_s) = P(f_t)$.

■

5.4 Robustness and inner rigidity

We now deduce some rigidity results about the mappings occurring in towers with bounded combinatorics and definite moduli.

Let $f : U \to V$ be infinitely renormalizable, and consider the Riemann surface $V - P(f)$ in its hyperbolic metric. For each $n \in \mathcal{SR}(f)$, $n > 1$, let $\gamma_n(f)$ denote the unique simple closed geodesic which is homotopic to a curve separating $J_n(f)$ from $P(f) - J_n(f)$. We say f is *robust* if

$$\liminf_{n \in \mathcal{SR}(f), n > 1} \ell(\gamma_n(f)) < \infty,$$

where $\ell(\cdot)$ denotes length in the hyperbolic metric on $V - P(f)$. Compare [Mc4, §9].

Theorem 5.8 (Robust rigidity) *If f is robust then the Julia set of f carries no invariant measurable line field.*

Proof. If f is robust then there is an annulus of definite modulus in the isotopy class of $\gamma_n(f)$ on $V - P(f)$, for infinitely many $n \in \mathcal{SR}(f)$. Moreover, a quasiconformal map distorts the modulus of an annulus by at most a bounded factor. Therefore the quadratic polynomial $g(z) = z^2 + c$ to which f is hybrid equivalent is also robust. By the main result of [Mc4], this robustness implies $J(g)$ carries no invariant line field. The latter property is also invariant under quasiconformal conjugacy, so $J(f)$ carries no invariant line field. ∎

Proposition 5.9 (Bounded lengths) *For any level $s > 1$ of a tower T in $\mathcal{T}ow^\infty(m, B)$, we have*

$$\ell(\gamma_s(f_1)) \leq L(m, B)$$

where $\ell(\cdot)$ denotes hyperbolic length on $V_1 - P(f_1)$.

Proof. By Proposition 5.7, there is an $\epsilon(m, B)$ such that for all $s > 1$, $W_s \cap P(f_1) = P(f_s)$, where W_s is an $\epsilon \cdot \mathrm{diam}(J(f_s))$-neighborhood of $J(f_s)$. Now there is an annulus of modulus m in V_1 surrounding

J_1, so by choosing ϵ still smaller if necessary (in terms of m) we can also assure that $W_s \subset V_1$ for all $s > 1$. By Proposition 4.8, there is an annulus of modulus at least $m(\epsilon) > 0$ surrounding $J(f_s)$ and lying in W_s. The core curve of this annulus is isotopic in $V_1 - P(f_1)$ to $\gamma_s(f_1)$, so by the Schwarz lemma $\ell(\gamma_s(f_1))$ is bounded above in terms of m and B.

∎

Corollary 5.10 (Inner rigidity) *For any level s of a tower $\mathcal{T}$ in $Tow^\infty(m, B)$, the mapping f_s is robust and therefore carries no invariant line field on its Julia set.*

Proof. By shift invariance, it suffices to prove the case $s = 1$, and this is contained in the preceding Proposition.

∎

Remarks. One can also give a *lower* bound on $\ell(\gamma_s(f_1))$; this point will be pursued further in §8.1. We will give a self-contained proof of the rigidity Corollary above in §8.3.

Next we check that $\mathrm{diam}(J_s)$ decreases at a definite rate as $s \to \infty$.

Corollary 5.11 *There is an $N(m, B)$ such that for any tower $\mathcal{T}$ in $Tow^\infty(m, B)$ and any pair of levels s, t with $t/s \geq N(m, B)$, we have*

$$\frac{\mathrm{diam}(J(f_t))}{\mathrm{diam}(J(f_s))} \leq \frac{1}{2}.$$

Proof. It suffices to prove the claim for $s = 1$ by shift invariance. Consider any tower $\mathcal{T}$ in $Tow^\infty(m, B)$. Since f_1 is robust, $\mathrm{diam}\, P(f_t) \to 0$ as $t \to \infty$ (compare [Mc4, Thm. 9.4]). But we have $\mathrm{diam}\, J(f_t) \leq C(m)\, \mathrm{diam}\, P(f_t)$ as an easy consequence of compactness of $Poly_2(m)$ [Mc4, Cor 5.10], so $\mathrm{diam}\, J(f_t) \to 0$. Thus there exists an N such that $\mathrm{diam}\, J(f_t)/\mathrm{diam}\, J(f_1) \leq 1/4$ for all $t \geq N$; and by continuity of $J(f_1)$ and $J(f_t)$, the ratio is less than $1/2$ on a neighborhood of $\mathcal{T}$. Thus we may choose N independent of $\mathcal{T}$ by compactness of $Tow^\infty(m, B)/\mathbb{C}^*$.

∎

Definition. The *full postcritical set* $P(T)$ of a tower T is given by

$$P(T) = \bigcup_{s \in S} P(f_s).$$

Corollary 5.12 *For any T in $Tow^\infty(m, B)$, the full postcritical set $P(T)$ is closed, as is $P(T) - P(f_s)$ for any $s \in S$.*

The sets $P(T)$ and $(P(T) - P(f_s))$ vary continuously with respect to T in the Hausdorff topology on closed subsets of $\mathbb{C}$.

Proof. If $\inf S = t$ is positive then $P(T) = P(f_t)$ and the Corollary follows easily.

Otherwise, as $t \to 0$, $P(f_t)$ is an increasing sequence of compact sets; to verify that $P(T)$ and $P(T) - P(f_s)$ are closed, we need only check that $d(P(f_t), P(f_u) - P(f_t)) \to \infty$ for adjacent levels $u < t$. But this distance is bounded below by $d(J(f_t), P(f_u) - P(f_t))$, which in turn is bounded below by $\epsilon(m, B) \cdot \operatorname{diam} J(f_t)$. Since $\operatorname{diam} J(f_t) \to \infty$, these sets are indeed closed.

Similarly, continuity follows from continuity of $P(f_t)$ for each t and the fact that $P(f_u) - P(f_t)$ is near ∞ for adjacent levels $u < t$ with t near zero.

∎

5.5 Unbranched renormalizations

A renormalization $f^n : U_n \to V_n$ of an iterate of $f \in Poly_2$ is *unbranched* if $V_n \cap P(f) = P_n(f)$. Similarly, a tower T is *unbranched* if $V_s \cap P(T) = P(f_s)$ for all $s \in S$.

A tower $T' = \langle g_s; s \in S' \rangle$ is a *restriction* of a tower $T = \langle f_s; s \in S \rangle$ if $S' \subset S$ and g_s is a restriction of f_s for each $s \in S'$. Then T' is a *subtower* of T. (Note: since each restricted map g_s is still quadratic-like, we have $J(g_s) = J(f_s)$ and $P(g_s) = P(f_s)$; compare [Mc4, Thm. 5.11].)

Proposition 5.13 *Any tower T in $Tow^\infty(m, B)$ can be restricted to an unbranched tower T' in $Tow^\infty(m', B)$ with $S' = S$, where $m' > 0$ depends only on (m, B).*

Proof. Let $\mathcal{T}$ be a tower in $Tow^{\infty}(m, B)$. For n large enough, $V_1' = f_1^{-n}(V_1)$ is close to $J(f_1)$, so it is a definite distance from $P(\mathcal{T}) - P(f_1)$. Then $f_1 : U_1' \to V_1'$ is an unbranched restriction of f_1, where $U_1' = f_1^{-1}(V_1')$. By continuity of $P(\mathcal{T}) - P(f_1)$, we have $V_1' \cap P(\mathcal{T}') = P(g_1)$ for all towers $\mathcal{T}' = \langle g_s; \ s \in S' \rangle$ near enough to $\mathcal{T}$. Thus we can use the same V_1' to construct unbranched renormalizations of g_1 with definite moduli for all towers $\mathcal{T}'$ sufficiently close to $\mathcal{T}$. By compactness of $Tow^{\infty}(m, B)/\mathbb{C}^*$, we find there is an $m'(m, B) > 0$ such that f_1 can always be restricted to an unbranched mapping of modulus at least m'. The same holds true for every level by shift invariance.

∎

6   Rigidity of towers

Let $T = \langle f_s : U_s \to V_s;\ s \in S \rangle$ and $T' = \langle g_s : U_s' \to V_s';\ s \in S \rangle$ be a pair of towers with the same level set S. A *conjugacy* ϕ between T and T' is a bijection

$$\phi : \bigcup V_s \to \bigcup V_s'$$

such that

$$\phi \circ f_s \ =\ g_s \circ \phi$$

for all $s \in S$. A conjugacy may be conformal, quasiconformal, smooth, etc. according to the quality of ϕ.

A tower T is *quasiconformally rigid* if any quasiconformal conjugacy ϕ from T to another tower T' is conformal.

Here are two equivalent formulations of quasiconformal rigidity which make no reference to T'. An L^∞ Beltrami differential $\mu = \mu(z)d\bar{z}/dz$ on $\bigcup V_s$ is *T-invariant* if $f_s^*(\mu) = \mu|U_s$ for all $s \in S$. Then we may assert:

A tower T is quasiconformally rigid if and only if any T-invariant Beltrami differential is zero (a.e.).

(By the measurable Riemann mapping theorem, invariant Beltrami differentials with $\|\mu\| < 1$ correspond bijectively to pairs (ϕ, T') modulo conformal conjugacy; compare [McS].)

By replacing μ with $\mu/|\mu|$ on the set where $\mu \neq 0$, it is clear that:

T is quasiconformally rigid if and only if T admits no invariant line field.

A tower with level set S is *bi-infinite* if $\inf S = 0$ and $\sup S = \infty$. Let $Tow_0^\infty(m, B)$ denote the set of bi-infinite towers with combinatorics bounded by B and with $\mathrm{mod}(U_s, V_s) \geq m$ for all s. Clearly $Tow_0^\infty(m, B)$ is closed in $Tow^\infty(m, B)$.

The main result of this section is:

Theorem 6.1 (Rigidity of towers) *A bi-infinite tower T with bounded combinatorics and definite moduli is quasiconformally rigid.*

(The hypothesis means $\mathcal{T} \in \mathit{Tow}_0^\infty(m, B)$ for some (m, B).)

Idea of the proof. In outline, the proof of rigidity follows the same lines as many other proofs in conformal dynamics. First, we may assume $\mathcal{T}$ is a *fine* tower (defined below), since this can be achieved by passing to a subtower. Given an invariant line field μ, pick a point z where $|\mu(z)| = 1$ and μ is almost continuous. Since there are no invariant line fields on the Julia sets $J(f_s)$, we see that $z \notin \bigcup J(f_s)$ and thus $f_s^k(z)$ eventually leaves the domain of f_s for every s. Next we use the fact that every f_s expands the hyperbolic metric on $\mathbb{C} - P(\mathcal{T})$, and that the expansion is by a definite factor as z is leaving the domain of f_s. Thus we can build up an arbitrary amount of expansion by choosing s sufficiently small, and blow up the nearly parallel line field near z to an almost holomorphic line field near $f_s^k(z)$. Passing to a limit (using compactness of towers), we construct a tower with a holomorphic invariant line field; this quickly leads to a contradiction. Thus the original tower admits no invariant line field.

A tower is *infinitely high* if $\inf S = 0$. We will also establish:

Theorem 6.2 (Relative rigidity) *An infinitely high tower with bounded combinatorics and definite moduli admits no invariant line field supported on $\mathbb{C} - \bigcup_S K(f_s)$.*

Finally we will show such a tower is determined up to isomorphism by the sequence of inner classes $I(f_s)$.

6.1   Fine towers

Although our results will hold for general towers with definite moduli and bounded combinatorics, it will be convenient to impose some additional geometric and combinatorial properties.

An infinitely renormalizable tower $\mathcal{T}$ is *fine* if

1. $\mathcal{T}$ is unbranched ($V_s \cap P(\mathcal{T}) = P(f_s)$ for all levels s);

2. $\mathcal{T}$ is nested ($V_t \subset U_s$ whenever $t > s$);

3. $\mathcal{T}$ has bounded combinatorics ($t/s \leq B$ for adjacent levels $t > s$);

4. T has definite moduli $(\text{mod}(U_s, V_s) \geq m > 0)$;

5. V_s and U_s are K-quasidisks; and

6. $\text{diam}(V_s) \leq C \, \text{diam} \, J(f_s)$ for all $s \in S$.

The constants (m, B, C, K) are independent of s; they are the *fine tower constants* of T. The set of all such fine towers will be denoted $Tow^\infty(m, B, C, K)$.

It is straightforward to verify:

Proposition 6.3 *The space $Tow^\infty(m, B, C, K)/\mathbb{C}^*$ of fine towers with specified bounds is compact.*

The closed subspace of bi-infinite fine towers will be denoted $Tow_0^\infty(m, B, C, K)$; it too is compact up to scaling.

Proposition 6.4 *Any tower T in $Tow^\infty(m, B)$ can be restricted to a fine tower T' in $Tow^\infty(m', B', C, K)$, where the fine tower constants depend only on (m, B). If T is bi-infinite then we can also arrange that T' is bi-infinite.*

Proof. By Proposition 5.13, we can pass to an unbranched tower with the same level set at the price of reducing m. By Proposition 4.10, for each $s \in S$, we may construct a further restriction $f_s : U_s' \to V_s'$ such that $\text{diam}(V_s') \leq C(m) \, \text{diam}(J(f_s))$, $\text{mod}(U_s', V_s') > m'(m)$, and both U_s' and V_s' are $K(m)$-quasidisks.

It remains to obtain the nesting condition; this will be achieved by restricting S. Since $J(f_s)$ is surrounded by an annulus of modulus at least $m'(m)/2$ in U_s', an $\epsilon(m) \cdot \text{diam} \, J(f_s)$-neighborhood of $J(f_s)$ is contained in U_s'. By Corollary 5.11, there is a constant $N(m, B)$ such that

$$\text{diam} \, J(f_t) < (\epsilon(m)/C(m)) \, \text{diam} \, J(f_s)$$

whenever $t \geq N(m, B)s$; this in turn implies V_t' lies in an $\epsilon(m)$-neighborhood of $J(f_s)$, and therefore $V_t' \subset U_s'$. So to obtain a nested subtower, we need only restrict to an $S' \subset S$ such that adjacent levels in S' are at least $N(m, B)$ apart. Let S' be a maximal such set containing 1. Since adjacent levels in S have ratio at most B, adjacent levels in S' have ratio at most $B' = N(m, B)B$. Also maximality guarantees $\sup S = \infty$, so the resulting tower

$$\mathcal{T}' = \langle f_s : U'_s \to V'_s; \; s \in S' \rangle$$

lies in $Tow^\infty(m', B', C, K)$. If $\mathcal{T}$ is bi-infinite then by maximality of S', $\mathcal{T}'$ is also bi-infinite.

∎

6.2 Expansion

For any tower $\mathcal{T}$, let

$$V(\mathcal{T}) = \bigcup_S V_s.$$

In this section we discuss expansion in the hyperbolic metric on $V(\mathcal{T}) - P(\mathcal{T})$.

Proposition 6.5 *For any $\mathcal{T}$ in $Tow_0^\infty(m, B)$, we have*

$$V(\mathcal{T}) = \mathbb{C}.$$

Proof. Since $\mod(U_s, V_s) \geq m > 0$, the region V_s contains an $\epsilon(m) \cdot \operatorname{diam}(J(f_s))$-neighborhood of $J(f_s)$. But $\operatorname{diam}(J(f_s)) \to \infty$ as $s \to 0$, so $\bigcup V_s = \mathbb{C}$.

∎

Lemma 6.6 *Let $U \subset \mathbb{H}$ be a connected region, and let $f : U \to \mathbb{H}$ be a holomorphic map such that $\|f'(z)\| \geq 1$ for all $z \in U$, with respect to the hyperbolic metric on $\mathbb{H}$. If $\|f'(z_0)\| = 1$ for some z_0, then f is an isometry.*

Proof. After composing with an isometry of $\mathbb{H}$, we can assume $f(z_0) = z_0$ and $f'(z_0) = 1$; we will show $f(z) = z$. Choose a small ball D about z_0 such that f^{-1} admits a univalent branch sending D to a neighborhood E of z_0. Since $\|(f^{-1})'\| \leq 1$, E is contained in D. But $f'(z_0) = 1$, so the Schwarz lemma implies $E = D$ and $f|E = \operatorname{id}$. Since U is connected, we conclude f is the identity throughout U, and in particular f is an isometry.

∎

Proposition 6.7 (Expansion) *Let T be a fine tower. Then every mapping f_s in T expands the hyperbolic metric on $V(T) - P(T)$.*

More precisely, if $z \in U_s$ and $f_s(z) \notin P(T)$, then $\|f_s'(z)\| > 1$ in the hyperbolic metric.

Proof. Let $W_s = V_s - P(f_s)$. Since T is unbranched and nested, we have $W_s \supset W_t$ whenever $s < t$. Let $Q(f_s) = f_s^{-1}(P(f_s))$. The mapping

$$f_s : (U_s - Q(f_s)) \to (V_s - P(f_s))$$

is a covering map, hence a local isometry between the respective hyperbolic metrics; and the proper inclusion

$$(U_s - Q(f_s)) \hookrightarrow (V_s - P(f_s))$$

is a contraction by the Schwarz lemma; so f_s expands the hyperbolic metric ρ_s on $V_s - P(f_s)$.

Since $f_t = f_s^{t/s}|U_t$ for every $t > s$, the map f_t also expands ρ_s for all $s < t$. If T has a minimal level s (that is, if $\inf S > 0$), then $V(T) - P(T) = V_s - P(f_s)$, so the Proposition is proved.

Now suppose $\inf S = 0$. Let ρ_0 denote the hyperbolic metric on $V(T) - P(T) = \mathbb{C} - P(T)$. As $s \to 0$, we have $\rho_s \to \rho_0$ uniformly on compact subsets of $V(T) - P(T)$. For any t and $z \in (U_t - Q(f_t))$, we have $\|f_t'(z)\| > 1$ in the ρ_s metric for all $s \le t$, so in the limit $\|f_t'(z)\| \ge 1$ in the ρ_0 metric.

It remains to show $\|f_t'(z)\| > 1$. If, on the contrary, $\|f_t'(z)\| = 1$, then by the preceding Lemma $f_t : (U_t - Q(f_t)) \to (V_t - P(f_t))$ is a local ρ_0-isometry. It follows that $Q(f_t) \subset P(T)$, so $Q(f_t) = P(f_t)$. Thus $P(f_t)$ is backward invariant by f_t, and therefore $P(f_t) = J(f_t)$. But this is impossible, because f_t is renormalizable.

∎

It also useful to know that the rate of expansion does not vary too rapidly:

Proposition 6.8 (Variation of expansion) *Let f_s^k be an interate of a mapping in a fine tower T. Let γ be a path in the domain of f_s^k joining x_1 to x_2, such that $f_s^k(\gamma)$ is disjoint from $P(T)$. Let*

d be the parameterized length of $f_s^k(\gamma)$ in the hyperbolic metric on $V(\mathcal{T}) - P(\mathcal{T})$. Then

$$\|(f_s^k)'(x_1)\|^\alpha \geq \|(f_s^k)'(x_2)\| \geq \|(f_s^k)'(x_1)\|^{1/\alpha},$$

where $\alpha = \exp(Md)$ for a universal $M > 0$.

Proof. For any $t < s$, we can write $f_s^k = f_t^n$ where $n = k(s/t)$. Let $W_t = f_t^{-n}(V_t - P(f_t))$. Then $f_t^n : W_t \to (V_t - P(f_t))$ is a covering map, so $\|f_t^n(z)\|_t = 1/\|\iota'(z)\|$, where the norm of the inclusion

$$\iota : W_t \hookrightarrow V_t - P(f_t)$$

is measured with respect to the hyperbolic metrics on domain and range, and $\|f_t^n(z)\|_t$ denotes the norm in the hyperbolic metric ρ_t on $V_t - P(f_t)$. By [Mc4, Cor 2.27], the distance between x_1 and x_2 in the hyperbolic metric on W_t controls the variation in $\|\iota'(x)\|$, and this distance is equal to the parameterized length of $f(\gamma)$ in the ρ_t metric. Thus the Proposition holds when d and $\|(f_s^k)'\|$ are measured using ρ_t. But ρ_t tends to the hyperbolic metric on $V(\mathcal{T}) - P(\mathcal{T})$ as $t \to \inf S$, so the result follows.

∎

Remark. Another way to formulate this theorem is that any branch of the multivalued-function

$$\log\log\left(\frac{1}{\|(f_s^{-k})'(z)\|}\right)$$

is uniformly Lipschitz in the hyperbolic metric on $V(\mathcal{T}) - P(\mathcal{T})$.

We now show that definite expansion is achieved as a point escapes. This is most conveniently proved for fine towers. Any point which is not in the filled Julia set of $f_s : U_s \to V_s$ eventually escapes from U_s under iteration. The last iterate occurs when z is in U_s but $f_s(z)$ is in $V_s - U_s$.

Proposition 6.9 (Definite expansion) *Let* T *be a fine tower.*
Then for f_s *in* T *and* $z \in U_s$ *such that* $f_s(z) \notin U_s$, *we have:*

$$\|f'_s(z)\| \geq \lambda > 1$$

with respect to the hyperbolic metric on $V(T) - P(T)$, *and*

$$d_s(z, J(f_s)) \leq D,$$

where d_s *is the hyperbolic metric on* $V_s - P(f_s)$.
Here λ *and* D *depend only on the fine tower constants of* T.

Proof. Suppose $T \in \mathit{Tow}^\infty(m, B, C, K)$. By shift invariance it suffices to prove the Proposition for $s = 1$. By hypothesis, $z \in f_1^{-1}(V_1 - U_1)$.

Since T is a fine tower, the annulus $\overline{V_1 - U_1}$ is bounded by K-quasicircles, and it varies continuously as a function of T in $\mathit{Tow}^\infty(m, B, C, K)$.[1] Similarly, $\overline{f_1^{-1}(V_1 - U_1)}$ varies continuously with T.

Let $Q(f_1) = f_1^{-1}(P(f_1))$. Then there is a constant D (depending only on the fine tower constants for T) such that

$$\sup\{d_1(\zeta, Q(f_1)) \ : \ \zeta \in \overline{f_1^{-1}(V_1 - U_1)}\} \leq D,$$

because the left hand side defines a continuous function on the compact set $\mathit{Tow}_0^\infty(m, B, C, K)/\mathbb{C}^*$. Since $Q(f_1) \subset J(f_1)$ we have established that $d_1(z, J(f_1)) \leq D$.

To complete the proof, we will show for any $z \in f_1^{-1}(V_1 - U_1)$,

$$\|f'_1(z)\| \geq \lambda = 1/C(D) > 1$$

in the hyperbolic metric on $V(T) - P(T)$. Here $C(\cdot)$ is the function appearing in Proposition 4.9 (Inclusion contraction).

First observe that f_1 is a restriction of $f_s^{1/s}$ for every $s < 1$. Let $Q'_s = f_s^{-1/s}(P(f_s))$, and let $U'_s = f_s^{-1/s}(V_s)$; since T is unbranched and nested, we have $U_1 \subset U'_s$ and $Q(f_1) = Q'_s \cap U_1$. The mapping

$$f_s^{1/s} : (U'_s - Q'_s) \rightarrow (V_s - P(f_s))$$

---

[1] The set $\overline{V_1 - U_1}$ does not vary continuously on $\mathit{Tow}_0^\infty(m, B)$; this is one reason we require quasidisks in fine towers. Compare §4.1.

is a covering map, hence a local isometry between the hyperbolic metrics on domain and range, while the inclusion

$$\iota_s : (U'_s - Q'_s) \to (V_s - P(f_s))$$

is a contraction by the Schwarz lemma. Since $f_s^{1/s} = f_1$ on U_1, for any $z \in U_1 - Q(f_1)$ we have

$$\|f_1'(z)\|_s = 1/\|\iota_s'(z)\|,$$

where the first norm is measured in the hyperbolic metric ρ_s on $V_s - P(f_s)$, and the second is measured between the metrics on the domain and range of ι.

By Proposition 4.9 (Inclusion contraction),

$$\|\iota_s'(z)\| \leq C(d_s(z, Q'_s)) < 1.$$

But $(V_1 - P(f_1)) \subset (V_s - P(f_s))$ and $Q(f_1) \subset Q'_s$, so the Schwarz lemma implies $d_s(z, Q'_s) \leq d_1(z, Q(f_1))$. Since $z \in f_1^{-1}(V_1 - U_1)$, we have $d_1(z, Q(f_1)) \leq D$, and thus

$$\|f_1'(z)\|_s \geq 1/C(D) > 1.$$

Finally ρ_s converges to the hyperbolic metric on $V(\mathcal{T}) - P(\mathcal{T})$ as $s \to \inf S$, so we obtain definite expansion in the hyperbolic metric on $V(\mathcal{T}) - P(\mathcal{T})$.

∎

6.3   Julia sets fill the plane

We now analyze the dynamics of a fine tower $\mathcal{T}$.

Suppose $z \in V(\mathcal{T})$; then $z \in V_s$ for some s, and we can consider the forward iterates $f_s^k(z)$. If z has an infinite forward orbit under f_s, then it belongs to $J(f_s)$, and we can restrict our attention to the single polynomial-like map f_s to understand the dynamics near z. If z has a finite orbit under f_s, it may still have an infinite forward orbit under f_t for some $t < s$, and so belong to $J(f_t)$.

The remaining possibility is that z does not belong to the Julia set of any mapping in $\mathcal{T}$. To try to develop an infinite forward orbit for z, we can proceed as follows. Pick s_1 such that $z_0 = z$ lies in U_{s_1}, and iterate until $z_1 = f_{s_1}^{k_1}(z_0)$ escapes from U_{s_1}. Then pick a level $s_2 < s_1$ so U_{s_2} contains z_1, and iterate until $z_2 = f_{s_2}^{k_2}$ escapes from U_{s_2}. In a bi-infinite nested tower, this construction can be prolonged indefinitely, and it is consistent: $f_t^{s/t}$ is an extension of f_s for any $t < s$, so the forward orbit under f_t contains the forward orbit under f_s. Even when $\inf S > 0$, it is useful to organize the forward orbit of z by levels in this way.

Proposition 6.10 *Given a fine tower $\mathcal{T}$ with levels $s > t$ and $z \in U_s - J(f_t)$, choose $k > 0$ so $f_t^k(z) \in U_t$ but $f_t^{k+1}(z) \notin U_t$. Then with respect to the hyperbolic metric on $V(\mathcal{T}) - P(\mathcal{T})$,*

$$\|(f_t^k)'(z)\| \geq (s/t)^\beta > 1.$$

Here $\beta > 0$ depends only on the fine tower constants for $\mathcal{T}$.

Proof. Let $t = s_n < s_{n-1} < \ldots < s_1 = s$ enumerate the levels between s and t. Since $z \notin J(f_t)$, z does not belong to $J(f_{s_i})$ for any of these levels. Thus the iterates of z under f_{s_1} eventually escape from U_{s_1}. By the nesting condition, $V_{s_1} \subset U_{s_2}$, so we can continue the forward orbit by iterating f_{s_2}. The orbit then escapes from U_{s_2}, and so on, until finally it reaches $f_t^{-1}(V_t - U_t)$.

By Theorem 6.9, each escape results in expansion by a definite factor $\lambda > 1$. By the chain rule, the expansion of the composition of these maximal iterates is at least λ^n. Furthermore the composition is equal to f_t^k, since each f_{s_i} is a renormalization of f_t. Since $s_i/s_{i+1} \leq B$ for some constant B, n is bounded below by a multiple of $\log(s/t)$. Thus $\lambda^n \geq (s/t)^\beta$, where $\beta > 0$ depends only on the fine tower constants. ∎

Proposition 6.11 (Asymptotic density of Julia sets) *For any fine tower $\mathcal{T}$, levels $s > t$ of $\mathcal{T}$ and $z \in U_s$, either $z \in J(f_t)$ or the distance*

$$d(z, J(f_t)) < D \left(\frac{t}{s}\right)^\eta$$

in the hyperbolic metric on $V(\mathcal{T}) - P(\mathcal{T})$. Here D and $\eta > 0$ depend only on the fine tower constants of $\mathcal{T}$.

Proof. Suppose $z \notin J(f_t)$. Choose $k > 0$ such that $w = f_t^k(z) \in f_t^{-1}(V_t - U_t)$. By the preceding Proposition, $\|(f_t^k)'(z)\| \geq M = (s/t)^\beta > 1$. By Proposition 6.9, there is a D depending only on the fine tower constants such that $d(w, J(f_t)) \leq D$ in the hyperbolic metric on $V_t - P(f_t)$. Let δ be a minimal geodesic in $V_t - P(f_t)$ joining w to a point p in $J(f_t)$, and let γ be a lift of δ by f_t^{-k} to a path with endpoints z and q.

By the Schwarz lemma, the length of δ is less than D in the hyperbolic metric on $V(\mathcal{T}) - P(\mathcal{T})$. By Proposition 6.8, $\|(f_t^k)'\|$ is bounded below by $M^{1/\alpha(D)}$ along γ, and thus the length of γ is less than $\epsilon = DM^{-1/\alpha(D)}$. But $J(f_t)$ is invariant under f_t, so we have constructed a point q in $J(f_t)$ within distance ϵ of z. Finally $\epsilon = D(t/s)^{\beta/\alpha(D)}$ so we obtain the Proposition with $\eta = \beta/\alpha(D)$.

∎

Corollary 6.12 (Julia sets fill the plane) *For any bi-infinite tower $\mathcal{T}$ in $\mathrm{Tow}_0^\infty(m, B)$, the union of the Julia sets $J(f_s)$ is dense in $\mathbb{C}$.*

Proof. Any $z \in \mathbb{C}$ belongs to U_s for some s. For $t < s$, either $z \in J(f_t)$ or the distance from z to $J(f_t)$ is bounded by the preceding result. As $t \to 0$ the bound tends to zero, so $\bigcup J(f_t)$ is dense.

∎

Remark. Proposition 6.11 has a nice formulation in terms of 3-dimensional hyperbolic geometry; see Theorem 8.8.

6.4 Proof of rigidity

A line field μ is *holomorphic* on an open set U if locally

$$\mu = \overline{\psi}/|\psi|,$$

where $\psi = \psi(z)dz^2$ is a holomorphic quadratic differential (not identically zero). A holomorphic line field on an open disk U is *univalent* if $\mu = h^*(d\overline{z}/dz)$ for some univalent map $h : U \to \mathbb{C}$.

The unit ball in $L^\infty(\widehat{\mathbb{C}}, d\bar{z}/dz)$ is compact in the weak* topology. If towers $\mathcal{T}_k \to \mathcal{T}$ admit invariant line fields μ_k, then any weak* limit μ of μ_k is $\mathcal{T}$-invariant. (The analogous statement for Kleinian groups is Proposition 2.8.) To prove rigidity, it will suffice to construct a limit μ which is holomorphic, in view of:

Lemma 6.13 *A tower $\mathcal{T}$ in $Tow_0^\infty(m, B)$ admits no invariant line field which is holomorphic on a nonempty open set.*

Proof. Let $U \subset \mathbb{C}$ be the maximal open set on which μ is holomorphic. If $f_s'(z) \neq 0$, then $z \in U$ if and only if $f_s(z) \in U$.

Suppose U is nonempty. Then by the density of $\bigcup J(f_s)$, U meets the Julia set of f_s for some s.

We claim $V_s \subset U$. First, since U meets the Julia set of f_s, we have $\bigcup f_s^k(U) = V_s$. Taking into account the condition on f_s', we can at least conclude $(V_s - P(f_s)) \subset U$. But $f_s|P(f_s)$ is injective, so every point $z \in P(f_s)$, other than the critical value v, is the image of a point in $V_s - P(f_s)$. Thus $V_s - \{v\} \subset U$. Finally $v \in U$ because $f_s(v) \in U$ and $f_s'(v) \neq 0$.

Since V_s is simply-connected, we can write $\mu = \bar{\psi}/|\psi|$ for some holomorphic quadratic differential ψ defined on V_s. By invariance of μ, we have $f_s^* \psi = A\psi$ for some $A > 0$. But then ψ has a zero at the critical point $z = 0$ of f_s, and all of its iterated preimages, which are dense in $J(f_s)$. Thus ψ vanishes identically, a contradiction.

■

Proof of Theorem 6.1 (Rigidity of towers). Suppose $\mathcal{T} = \langle f_s; s \in S \rangle$ is a bi-infinite tower with bounded combinatorics and definite moduli, admitting an invariant line field μ. We can assume $\mathcal{T}$ is a fine tower, say belonging to $Tow_0^\infty(m, B, C, K)$, since μ is invariant under any subtower of $\mathcal{T}$. Using expansion in the hyperbolic metric on $\mathbb{C} - P(\mathcal{T})$, we will promote μ to a holomorphic line field, and then obtain a contradiction by the preceding Lemma.

By Corollary 5.10, μ vanishes almost everywhere on $\bigcup J(f_s)$. Since $|\mu| = 1$ on a set of positive measure, we can choose $z \in \mathbb{C} - \bigcup J(f_s)$ such that $|\mu(z)| = 1$ and μ is almost continuous at z. This means for each $\epsilon > 0$,

$$\lim_{r \to 0} \frac{\text{area}\{w \in B(z, r) : |\mu(w) - \mu(z)| < \epsilon\}}{\text{area } B(z, r)} = 1.$$

For all $s \in S$ sufficiently small, we have $z \in U_s - J(f_s)$, so there is an integer $k(s)$ such that

$$z_s = f_s^{k(s)}(z) \in f_s^{-1}(V_s - U_s).$$

As s tends to zero, the norm of $(f_s^{k(s)})'(z)$ tends to infinity in the hyperbolic metric on $\mathbb{C} - P(T)$, by Proposition 6.10.

Let T_s denote T shifted by s and conjugated by

$$\alpha_s(z) = z / \operatorname{diam} J(f_s).$$

Then after passing to a subsequence of $s \to 0$, the towers T_s converge to a fine tower

$$T' = \langle g_s : U'_s \to V'_s; s \in S' \rangle.$$

We can also assume $\alpha_s(z_s) \to w$ for some $w \in A = \overline{g_1^{-1}(V'_1 - U'_1)}$. Here the closed annulus $A = \lim \alpha_s(f_s^{-1}(V_s - U_s))$ in the Hausdorff topology.

Passing to a further subsequence, we can assume $\mu_s = (\alpha_s)_*(\mu)$ converges weak* to a T'-invariant Beltrami differential μ'.

Choose an open disk $D \subset V'_1 - P(T')$ containing w. Then $\alpha_s \circ f_s^{k(s)}$ admits a univalent inverse h_s on D for all s sufficiently large. Since $\|(f_s^{k(s)})'(z)\| \to \infty$, the diameter of $h_s(D)$ tends to zero as $s \to 0$. Thus μ is very nearly parallel on $h_s(D)$, and by the Koebe distortion theorem, $\mu_s = h_s^*(\mu)$ is very nearly univalent on D. By compactness of univalent maps, $\mu'|D$ is exactly univalent. In particular, μ' is holomorphic on an open set, contradicting the preceding Lemma. Thus T has no invariant line field.

∎

Proof of Theorem 6.2 (Relative rigidity). Pick, as in the preceding proof, a point $z \in \mathbb{C} - \bigcup K(f_s)$ where the invariant line field μ is almost continuous. For any level s, the point z eventually escapes from U_s under iteration of f_s. As for bi-infinite towers, $\|f'_s\| \geq 1$ in the hyperbolic metric on $\mathbb{C} - P(T)$. So by pushing μ forward, for arbitrarily small s we can find a hyperbolic ball D_s of definite radius centered at a point in $f_s^{-1}(V_s - U_s)$, such that $\mu|D_s$ is nearly univalent.

Shifting by s and rescaling, we obtain towers $\mathcal{T}_s$ converging to a bi-infinite tower $\mathcal{T}'$ in $Tow_0^\infty(m, B)$ for a subsequence of s tending to zero. Passing to a further subsequence, we obtain a weak* limiting $\mathcal{T}'$-invariant Beltrami differential ν. There is a further subsequence such that D_s converges to a nontrivial disk D. The limit ν is nonzero, because it is univalent on D. This contradicts the rigidity of bi-infinite towers.

∎

6.5   A tower is determined by its inner classes

Given a rigidity result, such as that just proved for towers, it is desirable to find an invariant which classifies the rigid objects. In this section we show that for towers, such an invariant is provided by the inner classes $\langle I(f_s);\ s \in S \rangle$, a countable sequence of complex numbers.

Let $\mathcal{T} = \langle f_s;\ s \in S \rangle$ and $\mathcal{T}' = \langle g_s;\ s \in S' \rangle$ be a pair of towers. Conformal conjugacy is usually too fine an equivalence relation on the space of towers, because the domains U_s and V_s are not canonical. We say $\mathcal{T}$ is *isomorphic* to $\mathcal{T}'$ if $S = S'$ and if there is a conformal mapping ϕ between neighborhoods of $\bigcup K(f_s)$ and $\bigcup K(g_s)$ establishing a conjugacy between $f_s|K(f_s)$ and $g_s|K(g_s)$ for each s. Equivalently, two towers are isomorphic if they have restrictions (preserving the set of levels) which are conformally conjugate.

Theorem 6.14 (Inner invariants of towers) *Let* $\mathcal{T} = \langle f_s;\ s \in S \rangle$ *and* $\mathcal{T}' = \langle g_s;\ s \in S \rangle$ *be a pair of infinitely high towers with bounded combinatorics and definite moduli. Suppose* $I(f_s) = I(g_s)$ *for all* s. *Then* $\mathcal{T}$ *and* $\mathcal{T}'$ *are isomorphic.*

Proof.   We may assume both towers belong to $Tow(m, B)$. For any $s \in S$, $I(f_s) = I(g_s)$, so by Proposition 4.6 there is a K-quasiconformal hybrid conjugacy $\phi_s : V'_s \to V''_s$ between restrictions $f_s : U'_s \to V'_s$ and $g_s : U''_s \to V''_s$ with $\mathrm{mod}(U'_s, V'_s) > m' > 0$. Here K and m' depend only on m.

Since $\mathrm{diam}\, K(f_s) \to \infty$ as $s \to 0$, every compact set in $\mathbb{C}$ is eventually contained in V'_s. By compactness of normalized quasiconformal mappings, we can pass to a subsequence of $s \to 0$ such that $\phi_s \to \phi : \mathbb{C} \to \mathbb{C}$ uniformly on compact sets. The mapping ϕ conjugates $f_s|K(f_s)$ to $g_s|K(g_s)$ for every s, $\bar{\partial}\phi = 0$ a.e. on $\bigcup K(f_s)$, and ϕ conjugates $\mathcal{T}$ to another tower $\mathcal{T}''$. By the relative rigidity of towers, ϕ is conformal, so ϕ provides an isomorphism between $\mathcal{T}$ and $\mathcal{T}'$.

■

7 Fixed points of renormalization

In this chapter we study the convergence of iterated renormalization over the complex numbers.

We begin by defining the renormalization operators $\mathcal{R}_p$ and discussing their relation to tuning and the Mandelbrot set. Then we give a conjectural construction of fixed points of renormalization, starting from purely combinatorial data, namely a quadratic polynomial with a periodic critical point. The construction parallels the geometrization of 3-manifolds that fiber over the circle.

Next we justify several steps in the construction, starting with a quadratic polynomial f whose inner class is fixed under renormalization. We show that if the quadratic germs of $\mathcal{R}_p^n(f)$ have moduli bounded below, then they converge to a fixed point F of renormalization uniformly on a neighborhood of $J(F)$. Moreover F attracts all quadratic-like maps with the same inner class as f. The proof relies on the rigidity of towers.

To give a more precise discussion of the domain of convergence, we show F has a unique maximal analytic continuation $\tilde{F} : W \to \mathbb{C}$, and $\mathcal{R}_p^n(f)$ converges to $\tilde{F}$ uniformly on compact subsets of W.

Finally we use Sullivan's *a priori* bounds (Theorem 7.15) to complete the discussion in the setting of real mappings.

In §9.5 we will study the renormalization operators further, and show $\mathcal{R}_p^n(f) \to \tilde{F}$ exponentially fast.

7.1 Framework for the construction of fixed points

Let $\mathcal{H}$ denote the set of all holomorphic maps $f : W \to \mathbb{C}$ such that $W \subset \mathbb{C}$ is a topological disk, $f'(0) = 0$, and the restriction of f to some neighborhood of $z = 0$ is a quadratic-like map with connected Julia set.

We refer to $\mathcal{H}$ as the space of *extended quadratic-like maps*. It is topologized as follows: $f_i : W_i \to \mathbb{C}$ converges to $f : W \to \mathbb{C}$ if for every compact $K \subset W$, we have $K \subset W_i$ for all $i \gg 0$ and $f_i \to f$ uniformly on K. This topology is not Hausdorff; the closure of a point f includes all the restrictions of f that are still in $\mathcal{H}$.

Nevertheless it is a convenient topology for describing convergence of renormalization.

Lemma 7.1 *If* $f_i : U_i \to V_i$, $i = 1, 2$ *are two quadratic-like restrictions of* $f \in \mathcal{H}$ *to a neighborhood of* $z = 0$, *each with connected Julia set, then* $K(f_1) = K(f_2)$.

Proof. This result is an easy modification of [Mc4, Theorem 7.1].

First we claim $K(f_1) \cap K(f_2)$ is connected. Otherwise, there is a component U of $W - (K(f_1) \cup K(f_2))$ with $\overline{U}$ compact and $\partial U \subset K(f_1) \cup K(f_2)$. By forward invariance of the filled Julia sets and the maximum principle, the iterates $f^n | U$ are defined for all n and uniformly bounded. But $\overline{U}$ covers an interval in the ideal boundary of $K(f_i)$ for $i = 1$ or 2, so $f^n(U)$ contains an annulus surrounding $K(f_i)$, for some $n > 0$. Then $\{f^n\}$ is a normal family on $K(f_i)$, which is impossible (e.g. by the existence of repelling cycles).

Now let U and V be the components of $U_1 \cap U_2$ and $V_1 \cap V_2$ containing $K(f_1) \cap K(f_2)$. Then $f : U \to V$ is a polynomial-like map, and its degree is still two because the critical point $z = 0$ lies in U. This implies $K(f_1) = K(f_2) = K(f : U \to V)$.

∎

Julia set, modulus and normalization. By the Lemma, for $f \in \mathcal{H}$ we may unambiguously define $J(f)$ and $K(f)$ as the Julia set and filled Julia set of any quadratic-like restriction $f : U \to V$ with $0 \in U$ and with connected Julia set.

The *modulus* of f, denoted $\mathrm{mod}(f)$, is the supremum of $\mathrm{mod}(U, V)$ over all such quadratic-like restrictions.

Let us say $f \in \mathcal{H}$ is *normalized* if the β fixed point of $K(f)$ is at $z = 1$. Any quadratic-like map or quadratic polynomial with connected Julia set is affinely conjugate to a unique normalized map $f \in \mathcal{H}$. For example, for c in the Mandelbrot set, the normalized form of $P_c(z) = z^2 + c$ is $f(z) = \beta(c)z^2 + c/\beta(c)$, where

$$\beta(c) = \frac{1}{2} + \sqrt{\frac{1}{4} - c}$$

is the β-fixed point of P_c (with $\mathrm{Re}\,\beta(c) \geq 1/2$). A normalized real even mapping f restricts to a unimodal map $f : [-1, 1] \to [-1, 1]$, and $[-1, 1] = K(f) \cap \mathbb{R}$.

A variant of Proposition 4.3 is:

Proposition 7.2 *The set of normalized maps f with $\mathrm{mod}(f) \geq m > 0$ is sequentially compact.*

Quadratic germs. The space $\mathcal{G}$ of *quadratic germs* is the quotient of $\mathcal{H}$ by the equivalence relation $f_1 \sim f_2$ if $f_1|K(f_1) = f_2|K(f_2)$. Let $\pi : \mathcal{H} \to \mathcal{G}$ denote the projection of $f \in \mathcal{H}$ to its germ $[f] \in \mathcal{G}$.

We give $\mathcal{G}$ the following topology: $[f_n] \to [f]$ if and only if there are representatives such that $f_n \to f$ in $\mathcal{H}$. Note that convergence is required to take place on a definite neighborhood of $K(f)$, rather than on a neighborhood which can shrink as n increases. The space of germs is Hausdorff.

Every quadratic germ $[f]$ has a representative f_1 which is a quadratic-like mapping. We define the inner class $I(f)$ and the periods of simple renormalization $\mathcal{SR}(f)$ by $I(f) = I(f_1)$ and $\mathcal{SR}(f) = \mathcal{SR}(f_1)$. The functions $I(f)$ and $\mathcal{SR}(f)$ depend only on the germ of f.

The renormalization operator $\mathcal{R}_p$. For $p \geq 1$, let W_p denote the component of $f^{-p}(\mathbb{C})$ containing the origin. By the maximum principle, W_p is a disk. Let $\mathcal{H}^{(p)}$ denote the set of f such that $p \in \mathcal{SR}(f)$. Then there is a unique $\alpha \in \mathbb{C}^*$ such that $\alpha f^p(\alpha^{-1}z)$ is normalized.

The *renormalization operator*

$$\mathcal{R}_p : \mathcal{H}^{(p)} \to \mathcal{H}$$

is defined by

$$\mathcal{R}_p(f) = \alpha f^p \alpha^{-1} : \alpha W_p \to \mathbb{C}.$$

Let $\mathcal{G}^{(p)}$ be the image of $\mathcal{H}^{(p)}$; it consists of all germs $[f]$ that can be represented by quadratic-like maps f_1 such that f_1^p is simply renormalizable. It is easy to see $[\mathcal{R}_p(f)]$ depends only on $[f]$, so renormalization descends to a map

$$\mathcal{R}_p : \mathcal{G}^{(p)} \to \mathcal{G}.$$

Similarly the inner class $I(f)$ determines the inner class of $\mathcal{R}_p(f)$, so renormalization descends further to a mapping

$$\mathcal{R}_p : M^{(p)} \to M,$$

where M is the Mandelbrot set, and $M^{(p)}$ is the set of c such that $P_c^p(z)$ is simply renormalizable. Explicitly, this last map is given by $\mathcal{R}_p(c) = I(\mathcal{R}_p(P_c^p))$. Summing up, we obtain a commutative diagram:

$$
\begin{array}{ccc}
\mathcal{H}^{(p)} & \xrightarrow{\ \mathcal{R}_p\ } & \mathcal{H} \\[2pt]
{\scriptstyle \pi}\downarrow & & {\scriptstyle \pi}\downarrow \\[2pt]
\mathcal{G}^{(p)} & \xrightarrow{\ \mathcal{R}_p\ } & \mathcal{G} \\[2pt]
{\scriptstyle I}\downarrow & & {\scriptstyle I}\downarrow \\[2pt]
M^{(p)} & \xrightarrow{\ \mathcal{R}_p\ } & M.
\end{array}
$$

We can now state one of the central motivating problems in the subject:

> *Describe the fixed points of the renormalization operators* $\mathcal{R}_p$ *and the mappings they attract.*

The discussion of fixed points of $\mathcal{R}_p$ can be carried out at successively finer levels. A fixed point c in the Mandelbrot set corresponds to a quadratic map $f(z) = z^2 + c$ such that f^p is quasiconformally conjugate to f near $K(f)$, by a map which is conformal on the filled Julia set. A fixed point $[F]$ in the space of germs $\mathcal{G}$ is represented by a quadratic-like mapping F satisfying the Cvitanović-Feigenbaum functional equation

$$F(z) = \alpha F^p(\alpha^{-1} z)$$

for all $z \in K(F)$. The renormalization factor α satisfies $|\alpha| > 1$ since $J(F^p)$ is a proper subset of $J(F)$. Finally a fixed point $F : W \to \mathbb{C}$ in $\mathcal{H}$ satisfies the functional relation above throughout the domain of definition of either side of the equation.

A necessary condition for existence of a fixed point. Note that for any quadratic-like mapping $f \in \mathcal{H}$ such that $n \in \mathcal{SR}(f)$, the moduli $\operatorname{mod}(f, n)$ and $\operatorname{mod}(R_n(f))$ are equal.

Let F be a quadratic-like mapping whose germ $[F]$ is fixed by $\mathcal{R}_p$. Then the inner class $c = I(F)$ is also fixed by renormalization. From F we can build a tower $\mathcal{T}$ with level set $S = \{\ldots, p^{-2}, p^{-1}, 1, p, p^2, \ldots\}$ and $f_{p^n}(z) = \alpha^{-n} F(\alpha^n z)$. By Proposition 5.1, $\inf_n \mathrm{mod}(\mathcal{R}_p^n(F)) > 0$, so the same is true of any mapping hybrid equivalent to F. Thus we have:

Proposition 7.3 *If $\mathcal{R}_p([F]) = [F]$, and $I(F) = c$, then $f(z) = z^2 + c$ is hybrid equivalent to a simple renormalization of f^p, and $\inf_n \mathrm{mod}(f, p^n) > 0$.*

In the remainder of this section we discuss the conjectural availability of quadratic polynomials f satisfying these two necessary conditions. In the next section we will see these conditions are also *sufficient* for the existence of a fixed point $[F]$ with inner class c (Theorem 7.9).

The tuning invariant. A point c in the Mandelbrot set is *superstable* if $P_c^p(0) = 0$ for some $p \geq 1$; the least such p is the *period* of c, denote $\mathrm{per}(c)$. The point c is *primitive* if it is superstable, $\mathrm{per}(c) > 1$ and $\mathcal{SR}(P_c) = \{1, \mathrm{per}(c)\}$. The primitive points are the centers of the small cardioids in M.

Now suppose $f(z) = z^2 + a$ is infinitely renormalizable, and let $p > 1$ be the least integer $p > 1$ such that f^p is simply renormalizable. Then one may construct a natural primitive point $c = \Phi(a)$ of period p, such that P_c provides a combinatorial approximation to f. The mapping P_c is uniquely determined by the period p and the property that $P_c(z)$ is a combinatorial *quotient* of f. To construct P_c, one first collapses the small postcritical sets $P_p(1), \ldots, P_p(p)$ of f to single points. Then f determines a critically finite branched covering g on the sphere after collapsing. By a result of Thurston, g is combinatorially equivalent to a unique quadratic polynomial P_c. For more details, see [Mc4, §B.5].[1]

Now let $f \in \mathcal{H}$ be an infinitely renormalizable mapping with $\mathcal{SR}(f) = \{n_0, n_1, n_2, \ldots\}$. (We list the levels in ascending order

---

[1] *Correction:* On p.189 of [Mc4], the definition of *quotient* should require that ϕ_1 is a quotient map $\phi_1 : (S^2, f^{-1}P(f)) \to (S^2, g^{-1}P(g))$, homotopic among such maps to a representative of ϕ. The construction of g on pp.201-202 should be similarly modified.

with $n_0 = 1$.) The *tuning invariant*

$$\tau(f) = \langle c_0, c_1, \ldots \rangle$$

is the sequence of primitive points in M defined by $c_i = \Phi(I(\mathcal{R}_{n_i}(f)))$. We have $\mathrm{per}(c_i) = n_{i+1}/n_i$.

Conjecture 7.4 (Combinatorial rigidity) *An infinitely renormalizable quadratic polynomial $f(z) = z^2 + c$ is uniquely determined by its tuning invariant.*

This conjecture implies the density of hyperbolicity in the complex quadratic family (compare [Mc4, Thm. B.23]).

Tuning maps. For each superstable point s in the Mandelbrot set M, Douady and Hubbard have constructed a *tuning map* of M into itself; a detailed construction appears in [Dou1]. While a complete development of the theory of tuning has yet to appear, its main properties (both known and conjectural) are presented in [Mil].

Following Milnor, we will use $x \mapsto s * x$ to denote the tuning map determined by a superstable point $s \in M$. This map is characterized by the following conditions:

1. $s * 0 = s$;

2. $x \mapsto s * x$ is a homeomorphism of M into itself; and

3. if $x \neq 1/4$, then $s * x \in M^{(p)}$ and $\mathcal{R}_p(s * x) = x$, where $p = \mathrm{per}(s)$.

It is known that the star product is associative.

Conjecture 7.5 *For any sequence of primitive points $(c_0, c_1, c_2, \ldots)$ and any $x \in M$, the tuning*

$$c_0 * c_1 * \ldots * c_n * x$$

converges, as $n \to \infty$, to the unique $c \in M$ such that $\tau(P_c) = \langle c_0, c_1, \ldots \rangle$.

Renormalization simply shifts the tuning invariant: if $\mathcal{SR}(f) = \{n_0, n_1, \dots\}$ then

$$\tau(\mathcal{R}_{n_i}(f)) = \langle c_i, c_{i+1}, \dots \rangle.$$

Thus an inner class which is fixed by renormalization must have a periodic tuning invariant. If c is a superstable point of period p, the preceding conjecture implies the iterated product $c^{*n} * x$ converges to the unique point $c^\infty \in M$ such that $c * c^\infty = c^\infty$. Compare [Mil, Conjecture 1.1]. Thus for $p > 1$ we have:

Conjecture 7.6 (Classification of fixed points in M.) *The fixed points of $\mathcal{R}_p$ in the Mandelbrot set correspond bijectively to the superstable points $c \in M$ of period p, by the correspondence $c \mapsto c^\infty$.*

Conjecture 7.7 (Classification of fixed points in $\mathcal{G}$.) *For each $c \in M$ such that $\mathcal{R}_p(c) = c$, there is a unique fixed point $[F]$ of $\mathcal{R}_p$ in $\mathcal{G}$ with $I(F) = c$.*

A superstable c factors uniquely as a product $c_0 * \dots * c_{k-1}$ of primitive c_i's, and the tuning invariant of F should be obtained by periodically repeating $(c_0, \dots, c_{k-1})$.

Example. For the Feigenbaum polynomial $f(z) = z^2 + c$ (where $c = -1.401155\dots$), the tuning invariant $\tau(f) = \langle -1, -1, -1, \dots \rangle$, and $-1 * c = c$. It is at present unknown if c is the *unique* fixed point of tuning by -1.

In §7.3, we will see any fixed point $[F]$ for $\mathcal{R}_p$ in $\mathcal{G}$ has a unique lifting to a fixed point $\tilde{F}$ in $\mathcal{H}$.

Bounded moduli and combinatorics. Suppose $f \in \mathcal{P}oly_2$ is infinitely renormalizable, with $\mathcal{SR}(f) = \{n_0, n_1, n_2, \dots\}$. Then f has *bounded combinatorics*, or combinatorics *bounded by B*, if

$$\sup n_{i+1}/n_i \le B.$$

We say f has *definite moduli* if

$$\inf_{n \in \mathcal{SR}(f)} \operatorname{mod}(\mathcal{R}_n(f)) = m > 0.$$

Conjecture 7.8 (Definite moduli) *If $f(z) = z^2 + c$ is infinitely renormalizable with combinatorics bounded by B, then f has definite moduli; more precisely,*

$$\inf_{n \in \mathcal{SR}(f)} \mathrm{mod}(\mathcal{R}_n(f)) \geq m(B) > 0.$$

Now let $c \in M$ be a superstable point of period p. Here is a construction which, assuming the conjectures above, produces the fixed point of the renormalization operator $\mathcal{R}_p$ corresponding to c.

1. First, let c^∞ be the limit of the iterated tunings c^{*n}, and let $f(z) = z^2 + c^\infty$. Then $c * c^\infty = c^\infty$, so the inner class of $\mathcal{R}_p(f)$ is also c^∞. Therefore we have a quasiconformal conjugacy ϕ between f and $\mathcal{R}_p(f)$ near $K(f)$.

2. The map f has definite moduli at each level p^n of renormalization. Thus the iterates $[\mathcal{R}_p^n(f)]$ lie in a compact subset of the space of germs $\mathcal{G}$.

3. Let $\mathcal{T}$ be a tower with definite moduli built from the successive renormalizations of f; the level set of $\mathcal{T}$ is $S = \{p^n : n \geq 0\}$, and $f_1 = f$. Let $\mathcal{T}_n$ denote $\mathcal{T}$ shifted by p^n. Pass to a subsequence of n such that $\mathcal{T}_n \to \mathcal{T}_\infty$ and $[\mathcal{R}_p^n(f)] \to [F]$. Then $[F]$ is represented by the mapping F at level 1 in $\mathcal{T}_\infty$, and the union of the Julia sets in the tower $\mathcal{T}_\infty$ is dense in the plane.

4. The quasiconformal map ϕ gives rise to a limiting quasiconformal automorphism ψ of $\mathcal{T}_\infty$ shifting levels by p.

5. Since $\mathcal{T}_\infty$ is quasiconformally rigid, ψ is conformal; thus $\psi(z) = \alpha z$ with $|\alpha| > 1$.

6. Then $F^p(z) = \alpha^{-1} F(\alpha z)$ for all $z \in K_p(F)$, so $[F]$ is a fixed point of $\mathcal{R}_p$. By a similar argument, $[\mathcal{R}_p^n(g)] \to [F]$ for all g with $I(g) = c^\infty$.

7. By analytic continuation, $[F]$ lifts to a unique fixed point $\tilde{F} : W \to \mathbb{C}$ of $\mathcal{R}_p$ in $\mathcal{H}$.

This procedure is organized to parallel the construction of 3-manifolds which fiber over the circle given in §3.4.

In the next two sections we will justify steps (3-7) in the above construction, and in §7.4 we will see steps (1-2) at least work over the reals.

7.2 Convergence of renormalization

We now show rigidity of towers can be used to pass from compactness to convergence for renormalization. That is, if $f(z) = z^2 + c$, $\mathcal{R}_p(c) = c$, and $[\mathcal{R}_p^n(f)]$ ranges in a compact subset of $\mathcal{G}$, then the renormalizations converge to a fixed point of $\mathcal{R}_p$.

Theorem 7.9 (Convergence of renormalization) *Suppose that $f(z) = z^2 + c$ is hybrid conjugate to a simple renormalization of f^p, $p > 1$, and $\inf_n \operatorname{mod}(\mathcal{R}_p^n(f)) > 0$. Then there exists a unique fixed point $[F]$ of $\mathcal{R}_p$ in $\mathcal{G}$ with $I(F) = I(f)$.*

Moreover, $[\mathcal{R}_p^n(g)] \to [F]$ for all g with the same inner class as f. In particular, the renormalizations of $[f]$ converge to $[F]$ in $\mathcal{G}$.

Proof. Let $S = \{p^n \ : \ n \geq 0\}$, and choose renormalizations $f_s = f^s : U_s \to V_s$ for each $s \in S$, with $\operatorname{mod}(U_s, V_s) \geq m = \inf \operatorname{mod}(\mathcal{R}_p^n(f))$. Then $\mathcal{T} = \langle f_s; \ s \in S \rangle$ is a tower in $Tow^\infty(m, B)$ with $B = p$. By assumption $I(f_s) = I(f)$ for all s.

Let $\mathcal{T}_n$ denote $\mathcal{T}$ shifted by p^n and rescaled so $f_{1,n}$ is a normalized quadratic-like map (its β fixed point is at $z = 1$). Since $f_{1,n}$ has a definite modulus, after normalization $\operatorname{diam} J(f_{1,n}) \asymp 1$. By compactness of towers, we can pass to a subsequence such that $\mathcal{T}_n$ converges to a bi-infinite tower $\mathcal{T}_\infty = \langle g_s; \ s \in S_\infty \rangle$ in $Tow_0^\infty(m, B)$, where $S_\infty = \{p^n \ : \ -\infty < n < \infty\}$.

By continuity of the inner class, we have $I(g_s) = I(f)$ for all s. But by Theorem 6.14, the sequence of inner classes in $\mathcal{T}_\infty$ determines the germ of each mapping g_s. Since $\mathcal{T}_\infty$ and $\mathcal{T}_\infty$ shifted by p have the same inner classes, $[F] = [g_1] = [g_p] = [\mathcal{R}_p(F)]$ is a fixed point of renormalization.

Similarly, any two fixed points $[F_1]$ and $[F_2]$ with the same inner class as f give rise to towers $\mathcal{T}_1$ and $\mathcal{T}_2$ with the same inner invariants, so $[F_1] = [F_2]$ and thus the fixed point is unique.

If $I(g) = I(f)$, then $\inf \mathrm{mod}(\mathcal{R}_p^n(g))$ is also positive, so any accumulation point $[G]$ of $\mathcal{R}_p^n(g)$ can be embedded in a tower with the same inner invariants as $\mathcal{T}_\infty$. Thus $[G] = [F]$ and $[\mathcal{R}_p^n(g)] \to [F]$.

∎

7.3   Analytic continuation of the fixed point

In this section we show each fixed point of renormalization $[F]$ in the space of quadratic germs has a unique maximal analytic continuation $\widetilde{F} : W \to \mathbb{C}$. In addition, $\widetilde{F}$ is the unique lifting of $[F]$ to a fixed point of renormalization in $\mathcal{H}$, and $\mathcal{R}_p^n(f) \to \widetilde{F}$ for all $f \in \mathcal{H}$ with the same inner class as F.

The domain of $\widetilde{F}$ is an open dense subset of $\mathbb{C}$, containing $\alpha^n J(F)$ for all n. For example, when F is real, its analytic continuation $\widetilde{F}$ is defined along the entire real axis. Part of the graph of the analytic continuation of the Feigenbaum fixed point is shown in Figure 7.1.

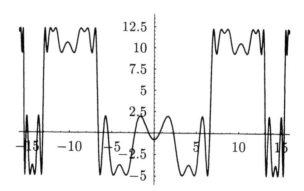

Figure 7.1. The analytic continuation of the Feigenbaum map.

Definitions. Let $f : U \to V$ and $g : U' \to V'$ be holomorphic maps between connected regions in $\mathbb{C}$. Say g is an *analytic continuation* of f if $f = g$ on some nonempty open set. If all such analytic continuations are restrictions of a single map $\widetilde{f}$, then $\widetilde{f}$ is the *unique maximal analytic continuation* of f.

Lemma 7.10 *Let* $f : U \to V$ *be a proper holomorphic map, and let* $g : U' \to V'$ *be an analytic continuation of* f. *If* $V' \subset V$, *then* $U' \subset U$ *and* $g = f|U$.

Proof. Let U'' be the union of the components of $U \cap U'$ on which $g = f$. By hypothesis U'' is open and nonempty, and by properness of f it is closed in U'.

∎

Theorem 7.11 *Let* $[F]$ *be a fixed point of the renormalization operator* $\mathcal{R}_p$. *That is, suppose* $F : U \to V$ *is a quadratic-like map satisfying*

$$F(z) = \alpha F^p(\alpha^{-1}z)$$

near $J(F)$, *where* $|\alpha| > 1$. *Then:*

1. *F has a unique maximal analytic continuation $\tilde{F} : W \to \mathbb{C}$;*

2. *the image of $\tilde{F}$ is $\mathbb{C}$;*

3. *the domain W of $\tilde{F}$ is an open, dense, simply-connected region in $\mathbb{C}$;*

4. *$\alpha^n J(F) \subset W$ for all n; and*

5. *$\tilde{F} = \lim \mathcal{R}_p^n(F)$ in $\mathcal{H}$.*

In fact, since $\mathcal{R}_p(F)$ is an analytic continuation of F, we have

$$\tilde{F}(z) \;=\; \mathcal{R}_p^n(F)(z) \;=\; \alpha^n F^{p^n}(\alpha^{-n}z)$$

wherever the right-hand side is defined.

Proof. For $n \geq 1$ let $F_n = \mathcal{R}_p^n(F)$. Then $F_n : W_n \to V_n$ is a polynomial-like map of degree 2^{p^n}, with $W_n = \alpha^n F^{-p^n}(V)$ and $V_n = \alpha^n V$. In particular, F_n is proper. By the functional equation, $F_n = F$ on a neighborhood of $J(F)$, so F_n is an analytic continuation of F.

Let $W = \bigcup W_n$, and let Z be the union of the domains of all analytic continuations of F. Clearly $W \subset Z$. Now consider compact connected set K with an analytic continuation g of F defined near K. Then $g(K) \subset \alpha^n V$ for all n sufficiently large (since V contains a neighborhood of the origin). By the Lemma, $K \subset W_n$ and $g = F_n | K$. Thus $Z = W$, and we have also shown that $\widetilde{F}(z) = \lim F_n(z)$ exists and agrees with $g(z)$ for any analytic continuation g of F. Since the definition of $\widetilde{F}$ is independent of g, $\widetilde{F}$ is the maximal analytic continuation of F.

We have also shown $\widetilde{F} = \lim \mathcal{R}_p^n(F)$, because any compact $K \subset W$ is eventually contained in W_n. Since each W_n is simply-connected, so is W.

The range of F_n is $\alpha^n V$, so the range of $\widetilde{F}$ is the whole plane. Similarly, the domain of $\widetilde{F}$ contains $\alpha^n J(F)$ for all $n > 0$, because $\alpha^n J(F) \subset W_n$.

Let

$$\mathcal{T} = \langle f_{p^n} = \alpha^{-n} F \alpha^n; n \in \mathbb{Z} \rangle$$

be the bi-infinite tower built from conjugates of F by α^n. Applying Corollary 6.12 to $\mathcal{T}$, we find $\bigcup \alpha^n J(F)$ is dense $\mathbb{C}$. This shows W is an open, dense, simply-connected subset of $\mathbb{C}$.

■

Corollary 7.12 *The map $\widetilde{F} : W \to \mathbb{C}$ is the unique fixed point of $\mathcal{R}_p$ in $\mathcal{H}$ with germ $[F]$.*

Proof. To see $\widetilde{F}$ is a fixed point, we need to check that $\widetilde{F}$ and $\mathcal{R}_p(\widetilde{F})$ have the same domain of definition. One inclusion follows from the fact that $\widetilde{F}$ admits no further analytic continuation. For the other, note that any compact set K in the domain of $\widetilde{F}$ is also contained in the domain of $\mathcal{R}_p^{n+1}(F)$ for all $n \gg 0$, so it is in the domain of $\mathcal{R}_p(\lim \mathcal{R}_p^n(F)) = \mathcal{R}_p(\widetilde{F})$.

If G is another fixed point with the same germ as F, then $G = \mathcal{R}_p^n(G) \to \widetilde{F}$, so the domain of G contains the domain of $\widetilde{F}$; but then $G = \widetilde{F}$ by maximality of the analytic continuation.

■

Theorem 7.13 *Let $\widetilde{F}$ be a fixed point of $\mathcal{R}_p$ in $\mathcal{H}$. Then $\mathcal{R}_p^n(f) \to \widetilde{F}$ for any $f \in \mathcal{H}$ with the same inner class as $\widetilde{F}$.*

Proof. It suffices to prove the result for a quadratic-like restriction of f. Since $I(f) = I(\widetilde{F})$, we can assume there is a quasiconformal map $\phi : \mathbb{C} \to \mathbb{C}$ conjugating f to a quadratic-like restriction F of $\widetilde{F}$.

From ϕ we obtain quasiconformal maps ϕ_n conjugating $\mathcal{R}_p^n(f)$ to $\mathcal{R}_p^n(F)$. Since the critical point and the β fixed point are preserved by ϕ_n, each conjugacy fixes $z = 0$ and $z = 1$. Any limit of ϕ_n gives a quasiconformal conjugacy to the tower generated by F, so by rigidity of towers ϕ_n converges to the identity uniformly on compact sets. Since ϕ_n conjugates $\mathcal{R}_p^n(f)$ to $\mathcal{R}_p^n(F)$, and $\mathcal{R}_p^n(F) \to \widetilde{F}$, we have $\mathcal{R}_p^n(f) \to \widetilde{F}$.

∎

Remarks. The mapping $\widetilde{F} : W \to \mathbb{C}$ is an infinite-sheeted *branched covering*, in the sense that each $z \in \mathbb{C}$ has a neighborhood V such $\widetilde{F} : U \to V$ is a proper map for each component U of $\widetilde{F}^{-1}(V)$. Moreover $\widetilde{F}$ is σ-*proper*, that is, its graph is a countable union of graphs of proper mappings, namely those of $\mathcal{R}_p^n(F) \in \mathcal{H}$.

The critical values of $\widetilde{F}$ are discrete. For example, when F is the Feigenbaum mapping, the critical values of $\widetilde{F}$ are certain endpoints of intervals complementary to the "large Cantor set"

$$P(\mathcal{T}) = \bigcup_{n \geq 0} \alpha^n P(F) \subset \mathbb{R},$$

the postcritical set of the tower generated by F. In fact, the dynamical system $(P(\mathcal{T}), \widetilde{F})$ is topologically conjugate to $(\mathbb{Q}_2, x \mapsto x + 1)$, where $\mathbb{Q}_2$ denotes the 2-adic completion of the rationals. The conjugacy can be chosen uniquely so $0 \in \mathbb{Q}_2$ corresponds to the critical point of F; then $P(F)$ corresponds to $\mathbb{Z}_2$, and the critical values of $\widetilde{F}$ correspond to the dyadic rationals $\{k/2^n \ : \ 0 < k \leq 2^n, n = 0, 1, 2, \dots\}$ (which are discrete in $\mathbb{Q}_2$).

The arguments of this section can also be adapted to show:

Theorem 7.14 *Let $\mathcal{T} = \langle f_s; s \in S \rangle$ be an infinitely high tower with bounded combinatorics and definite moduli. Then f_1 admits a unique maximal analytic continuation $\widetilde{f_1} : W \to \mathbb{C}$, defined by*

$$\widetilde{f_1}(z) = \lim_{s \to 0} f_s^{1/s}(z).$$

The analytic continuation is a surjective, σ-proper branched covering of the plane, W is simply-connected and dense in $\mathbb{C}$, and $J(f_s) \subset W$ for all $s \in S$.

7.4   Real quadratic mappings

In this section we combine Sullivan's *a priori* bounds with the rigidity of towers to construct fixed points of renormalization for *real* quadratic mappings. We also deduce the existence and uniqueness of bi-infinite real towers with prescribed bounded combinatorics. These results were first established in [Sul5].

The tuning invariant of a real quadratic polynomial is equivalent to the sequence $(\sigma_0, \sigma_1, \dots)$ of *shuffle permutations* of [Sul5] or the *unimodal, non-renormalizable permutations* of [MeSt]. The existence of mappings with given (real) tuning invariants is known quite generally for full families of unimodal maps, as a consequence of the kneading theory [MeSt, Prop. VI.1.3]. In particular any real tuning invariant is realized by a real quadratic polynomial.

For real tuning invariants with bounded combinatorics, Sullivan shows this real polynomial is unique [Sul5, Thm. 6]. The construction of renormalizations with definite moduli is carried out in [Sul5, §8]. In summary, we have the following *a priori* bounds:

Theorem 7.15 (Sullivan) *An infinitely renormalizable real quadratic polynomial $f(z) = z^2 + c$ with combinatorics bounded by B has definite moduli $m(B)$ at all levels of renormalization, and f is uniquely determined by its tuning invariant.*

We may now establish:

Theorem 7.16 (Real fixed points of renormalization) *Let $c \in \mathbb{R}$ be a superstable point of period $p > 1$. Then:*

1. *the iterated tunings c^{*n} converge to a point $c^\infty \in M$;*

2. *there is a unique quadratic-like germ $[F]$ fixed by $\mathcal{R}_p$ with inner class c^∞; and*

3. *for any quadratic-like map g with $I(g) = c^\infty$, we have*

$$\mathcal{R}_p^n(g) \to \tilde{F}$$

uniformly on compact subsets of W, where $\tilde{F} : W \to \mathbb{C}$ is the maximal analytic continuation of F.

In particular, $\mathcal{R}_p^n(f) \to \tilde{F}$ when $f(z) = z^2 + c^\infty$.

Proof. Let $c_0 * \ldots * c_{k-1}$ be the factorization of c into primitives. Then for any limit d of c^{*n}, the tuning invariant $\tau(z^2 + d)$ is the periodic sequence $\langle c_0, \ldots, c_{k-1}, c_0, \ldots, c_{k-1}, \ldots \rangle$. By Theorem 7.15, $z^2 + d$ is uniquely determined by its tuning invariant, so all limits agree and c^{*n} converges. (Alternatively, one can use the kneading invariant and the monotonicity theorem [MeSt, §II.10] to show c^{*n} is a monotone decreasing sequence of real numbers.)

Let $c^\infty = \lim c^{*n}$. Theorem 7.15 also guarantees that all renormalizations of $f(z) = z^2 + c^\infty$ have definite moduli. Since $\tau(\mathcal{R}_p(f)) = \tau(f)$, the map f is hybrid conjugate to $\mathcal{R}_p(f)$. By Theorem 7.9, $[\mathcal{R}_p^n(f)]$ converges to the unique fixed point $[F]$ of $\mathcal{R}_p$ with $I(F) = I(f)$. The same is true for any g with the same inner class, and Theorem 7.13 gives convergence of $\mathcal{R}_p^n(g)$ to the maximal analytic continuation $\tilde{F}$. ∎

Finally we restate and reprove [Sul5, Thm. 2] in the language of towers:

Theorem 7.17 *Given any bi-infinite sequence $\langle \ldots c_{-1}, c_0, c_1, \ldots \rangle$ of real primitive points with $\mathrm{per}(c_i)$ bounded, there exists a tower*

$$\mathcal{T} = \langle f_s; \, s \in S = \{ \ldots s_{-1}, s_0 = 1, s_1, \ldots \} \rangle$$

with definite moduli at all levels such that $s_{i+1}/s_i = \mathrm{per}(c_i)$ and f_{s_i} has tuning invariant $\langle c_i, c_{i+1}, \ldots \rangle$.

Any two such towers are isomorphic.

Proof. Let $B = \sup \mathrm{per}(c_i)$. First suppose the sequence c_i is periodic; that is $c_{i+k} = c_i$ for some $k > 0$. Let

$$p = \mathrm{per}(c_0) \, \mathrm{per}(c_1) \cdots \mathrm{per}(c_{k-1}).$$

There is a unique real quadratic polynomial $f(z) = z^2 + c$ with tuning invariant $\langle c_0, c_1, \ldots \rangle$. By Sullivan's *a priori* bounds we may construct a tower $T \in Tow^{\infty}(m(B), B)$ with $f_1 = f$ and $S = \mathcal{SR}(f) = \{s_0 = 1, s_1, s_2, \dots \}$; here $s_i = \prod_0^{i-1} \text{per}(c_j)$. Then $\tau(f_{s_i}) = \langle c_i, c_{i+1}, \dots \rangle$.

Let T_n denote T shifted by p^n and scaled so $\text{diam}(J(f_{1,n})) = 1$. By periodicity of the tuning invariant, the inner class of $f_{s,n}$ is independent of n for all $s > p^{-n}$. Passing to a convergent subsequence, we obtain a limiting tower T_∞ in $Tow_0^{\infty}(m(B), B)$ with the required tuning invariants at each level.

To obtain a tower with an arbitrary bi-infinite tuning invariant, approximate by a periodic one, and pass to a limit using compactness of towers. (This step uses the fact that $m(B)$ depends only on B.)

The bi-infinite sequence $\langle \dots, c_{-1}, c_0, c_1, \dots \rangle$ determines the inner class $I(f_s)$ for each f_s in the tower. Thus uniqueness up to isomorphism is immediate from Theorem 6.14.

$\blacksquare$

Notes and references. Sullivan's proof uses a different argument, based on Riemann surface laminations, to conclude uniqueness. The formulation of [Sul5, Thm. 2] was motivated by the tower approach, presented at IHES in 1990.

The density of hyperbolicity in the real quadratic family (announced in [Sw], [Lyu]) implies *any* infinitely renormalizable real quadratic polynomial is determined up to conformal conjugacy by its tuning invariant. The corresponding statement for complex quadratic polynomials is open. In fact, one does not yet know if any *complex* polynomial g with the same tuning invariant as the Feigenbaum polynomial f is conformally conjugate to f.

If Theorem 7.15 can be established for complex mappings, the results of this section will immediately generalize to arbitrary bounded combinatorics. Progress in this direction appears in [Lyu].

8 Asymptotic structure in the Julia set

Let $f(z) = z^2 + c$ be an infinitely renormalizable quadratic polynomial with bounded combinatorics and definite moduli. That is, suppose there are constants B and $m > 0$ such that $\mathcal{SR}(f) = \{n_0, n_1, n_2, \dots\}$ satisfies $\sup n_{i+1}/n_i \leq B$ for all i, and $\mathrm{mod}(f, n) \geq m > 0$ for all $n \in \mathcal{SR}(f)$.

By Sullivan's *a priori* bounds, any *real* infinitely renormalizable quadratic polynomial with bounded combinatorics automatically has definite moduli. In particular, the Feigenbaum polynomial is an example of such an f.

In this section we collect together some additional results about the geometry and dynamics of these polynomials. We show the complement of the postcritical set is a hyperbolic surface of bounded geometry (§8.1); the critical point is a deep point of the Julia set (§8.2); and at every scale near a point in $J(f)$, the full dynamics of f includes a quadratic-like map (§8.3). A more general theory of towers is sketched in §8.4.

We will write $A \asymp B$ to mean that $C_1 A < B < C_2 A$ for positive constants C_1 and C_2 depending only on (m, B).

8.1 Rigidity and the postcritical Cantor set

Recall there are natural simple closed curves γ_n surrounding the small postcritical sets $P_n = P(f) \cap J_n(f)$ for all $n \in \mathcal{SR}(f)$, $n > 1$. These simple curves are geodesics on the hyperbolic Riemann surface $\mathbb{C} - P(f)$. By definition, robustness of f means $\liminf \ell(\gamma_n) < \infty$. In this section we will obtain more precise geometric information about lengths in $\mathbb{C} - P(f)$; in fact we will determine the Riemann surface $\mathbb{C} - P(f)$ up to quasi-isometry.

Proposition 8.1 *Let $f(z) = z^2 + c$ have bounded combinatorics and definite moduli. Then for all $n \in \mathcal{SR}(f)$:*

1. *$\mathrm{diam}\, P_n(i) \asymp \mathrm{diam}\, J_n(i)$, for $1 \leq i \leq n$.*

2. *$n^{-\alpha} \leq \mathrm{diam}\, J_n \leq 4n^{-\beta}$ for constants $\alpha, \beta > 0$ depending only on (m, B).*

3. *There is an* unbranched *renormalization* $f^n : U_n \to V_n$ *such that* $\mathrm{mod}(U_n, V_n) \geq m'(m, B) > 0$.

4. $\ell(\gamma_n) \asymp 1$, *for* $n > 1$; *in particular,* f *is robust.*

Proof. (1) The diameters of $P_n(i)$ and $J_n(i)$ are comparable with constants depending only on m by [Mc4, Cor 5.10]. (For a quadratic polynomial g, if $\mathrm{diam}\, P(g) \ll \mathrm{diam}\, J(g)$ then g has an attracting fixed point. The same result holds for quadratic-like maps $g : U \to V$ with a constant depending on $\mathrm{mod}(U, V)$.)

(2) Since f has bounded combinatorics and definite moduli, we may construct a tower $\mathcal{T}$ in $\mathcal{T}ow^\infty(m, B)$ with $f_1 = f$ and level set $S = \mathcal{SR}(f)$. Then the diameters of the Julia sets shrink at a definite rate by Corollary 5.11, but not too quickly by Proposition 4.13. (Note that $2 < \mathrm{diam}\, J(f) \leq 4$ for any c in the Mandelbrot set.)

(3) By Propositions 5.13 and 5.1, we can obtain a subtower with the same level set S such that each f_s is an unbranched renormalization of f_1 with modulus greater than $m'(m, B) > 0$.

(4) For any unbranched renormalization $f^n : U_n \to V_n$, with $\mathrm{mod}(U_n, V_n) > m'$, the annulus $V_n - U_n$ represents the same homotopy class as γ_n on $\mathbb{C} - P(f)$, so we have an upper bound on the hyperbolic length of $\ell(\gamma_n)$. For the lower bound, observe that if $\ell(\gamma_n)$ is very short, then the Euclidean diameter of P_n is much less than that of $P_{n'}$, where $n' < n$ is the adjacent lower level of simple renormalization. This implies $\mathrm{diam}\, J_n \ll \mathrm{diam}\, J_{n'}$ by (1), which contradicts Proposition 4.13 when $\ell(\gamma_n)$ is sufficiently small.

■

Since robustness implies rigidity [Mc4, Thm. 1.7], we have:

Corollary 8.2 *Any polynomial quasiconformally conjugate to* f *is conformally conjugate to* f.

A self-contained proof of this Corollary will be given in §8.3.

Corollary 8.3 (Hu-Jiang) *The Julia set of* f *is locally connected.*

See [JH], [Ji]; these authors show in the setting of bounded combinatorics, local connectivity follows from the existence of *unbranched* renormalizations of definite modulus, and we have just shown these exist.

For $1 \leq i \leq n$ let $\gamma_n(i)$ be the geodesic representing the isotopy class of a curve separating $J_n(i)$ from $P(f) - J_n(i)$. The curve $\gamma_n(i)$ encloses the small postcritical set $P_n(i) = f^i(P_n)$, and $\bigcup P_n(i) = P(f)$. Note $\gamma_n(n) = \gamma_n$ encloses P_n.

Consider the collection

$$\Gamma = \{\gamma_n(i) \ : \ n \in \mathcal{SR}(f), n > 1, 1 \leq i \leq n\}$$

of all such geodesics. These curves are pairwise disjoint, and they partition the Riemann surface $\mathbb{C} - P(f)$ into planar pieces, each with at most $B + 1$ boundary components (see Figure 8.1). There is a unique piece containing a neighborhood of infinity in $\mathbb{C}$, which is geometrically a cusp; all other pieces are bounded by closed geodesics.

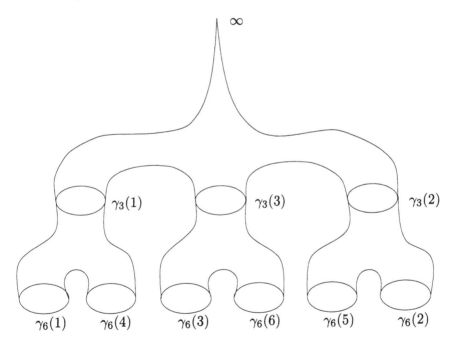

Figure 8.1. Part of the Riemann surface $\mathbb{C} - P(f)$.

Theorem 8.4 (Bounded geometry) *All curves in Γ have length bounded above and below in terms of (m, B).*

Outside a standard neighborhood of the cusp at infinity, the Riemann surface $\mathbb{C} - P(f)$ has injectivity radius bounded above and below in terms of (m, B).

Proof. By the preceding result, we have $\ell(\gamma_n) \asymp 1$ uniformly in n. But all the curves $\gamma_n(i)$ at a given level n have comparable lengths; in fact

$$\frac{1}{2}\ell(\gamma_n) \leq \ell(\gamma_n(i)) \leq \ell(\gamma_n)$$

for $i = 1, 2, \ldots, n$ [Mc4, Thm. 9.3]. Thus we have upper and lower bounds for the lengths of all the curves in Γ.

Next we verify a lower bound on the injectivity radius. Suppose δ is a short geodesic on $\mathbb{C} - P(f)$. If δ is disjoint from Γ, there is a curve $\gamma_n(i)$ enclosed by δ with n minimal (an outermost curve in the disk bounded by δ in $\mathbb{C}$). If δ is very short, then $J_n(i)$ is much smaller than the Julia set containing it at the adjacent lower level of renormalization, contradicting Proposition 4.13.

Now suppose δ intersects some curve γ in Γ. The δ cannot be short because an upper bound on $\ell(\gamma)$ gives a lower bound on the width of a standard collar about γ, which δ must traverse.

Thus the injectivity radius in $\mathbb{C} - P(f)$ is bounded below. For the upper bound, cut along the curves in Γ; we obtain pieces with bounded area. Every point in a given piece is near a boundary geodesic of bounded length, so the injectivity radius is bounded above.

■

From the lower bound on the injectivity radius we have (see e.g. [Mc4, Thm. 2.3]):

Corollary 8.5 *Away from the cusp at $z = \infty$, the hyperbolic metric $\rho(z)|dz|$ on $\mathbb{C} - P(f)$ is comparable to the $1/d$ metric; that is, $\rho(z) \asymp 1/d(z, P(f))$ where d denotes Euclidean distance.*

Corollary 8.6 *The postcritical set $P(f)$ is quasiconformally equivalent to the Cantor middle-thirds set. In particular, its Hausdorff dimension is less than two.*

Proof. Complete the system of curves Γ to a maximal system of disjoint curves Γ' on the Riemann surface $X = \mathbb{C} - P(f)$. By the bounds obtained above, the curves appearing in Γ' can also be chosen to have length bounded above and below. Cutting along the curves Γ', we obtain a decomposition of X into pairs of pants, each with bounded geometry (apart from the one component containing the cusp at infinity). There is a combinatorially identical bounded pair of pants decomposition for the complement $Y = \mathbb{C} - C$ of the standard middle-thirds Cantor set C. Thus we may construct a quasiconformal map $\phi : X \to Y$.

By robustness, $P(f)$ has absolute area zero [Mc4, Thm. 9.4], and thus ϕ extends to a quasiconformal map of the sphere to itself sending $P(f)$ to C. This shows $P(f)$ is quasiconformally equivalent to the middle-thirds Cantor set.

Sets of Hausdorff dimension two are preserved under quasiconformal maps [GV], so $P(f)$ also has dimension less than two.

∎

Remark. One may show more precisely that there is a quasiconformal map $\phi : \mathbb{C} \to \mathbb{C}$ sending $P(f)$ to a standard model for the Cantor set $\operatorname{proj\,lim}_{n \in \mathcal{SR}(f)} \mathbb{Z}/n$ and conjugating $f : P(f) \to P(f)$ to $x \mapsto x + 1$.

8.2 Deep points of Julia sets

Milnor has conjectured that if one magnifies the Mandelbrot set around the fixed point s^∞ of a tuning map $x \mapsto s * x$, then the magnified sets become very nearly dense in the plane [Mil, Conj. 1.3].

It is a metatheorem that the structure of the Mandelbrot set at c mimics the structure of the Julia set of $z^2 + c$ at its critical value c. (See, for example, [Tan].) In this section we give a result supporting Milnor's conjecture, but stated for Julia sets rather than for the Mandelbrot set.

Theorem 8.7 *Let $f(z) = z^2 + c$ be infinitely renormalizable, with bounded combinatorics and definite moduli. Then the Julia set $J(f)$*

converges to the whole plane when it is magnified about the critical point $z = 0$.

Proof. Any Hausdorff limit of the magnified Julia sets $\lambda \cdot J(f)$, $\lambda \to \infty$, contains the union of all of the Julia sets in a bi-infinite tower constructed from renormalizations of f. This tower has bounded combinatorics and definite moduli, so the union of its Julia sets is dense in the plane (Corollary 6.12).

∎

Here is a more quantitative version formulated in the terminology of §2.5.

Theorem 8.8 (Deep critical points) *The critical point $z = 0$ is a deep point of the Julia set $J(f)$. More precisely, let $\gamma : [0, \infty) \to \mathbb{H}^3 \cong \mathbb{C} \times \mathbb{R}_+$ be the geodesic ray $\gamma(\tau) = (0, e^{-\tau})$. Then there is an $\epsilon(m, B) > 0$ such that*

$$d(\gamma(\tau), \partial \operatorname{hull}(J)) > \epsilon\tau > 0$$

for all $\tau > 0$.

Lemma 8.9 *For $0 < \epsilon < 1$ the cone of consisting of all points within distance $\epsilon\tau$ of $\gamma(\tau)$ for some $\tau > 0$ lies above the graph of the function $t = |z|^\alpha$, where $\alpha = (1 + \epsilon)/(1 - \epsilon)$.*

Proof. Let $t_0 = e^{-\tau}$. The hyperbolic ball B about $(0, t_0)$ of radius $\epsilon\tau$ is the same as the Euclidean ball in $\mathbb{C} \times \mathbb{R}_+$ which has as its diameter the segment $[t_0^{1+\epsilon}, t_0^{1-\epsilon}]$ along the t-axis. If (z, t) lies in this ball, then $|z| < t_0^{1-\epsilon}$ and $t > t_0^{1+\epsilon}$, so $t > |z|^\alpha$.

∎

Proof of Theorem 8.8. First note that γ is contained in the convex hull of the Julia set. Indeed, $J(f)$ meets the unit circle in a nonempty set invariant under $z \mapsto -z$, so $\gamma(0) = (0, 1) \in \operatorname{hull}(J(f))$; and the rest of γ lies in $\operatorname{hull}(J(f))$ because the critical point $(0, 0) = \lim_{\tau \to \infty} \gamma(\tau)$ lies in $J(f)$.

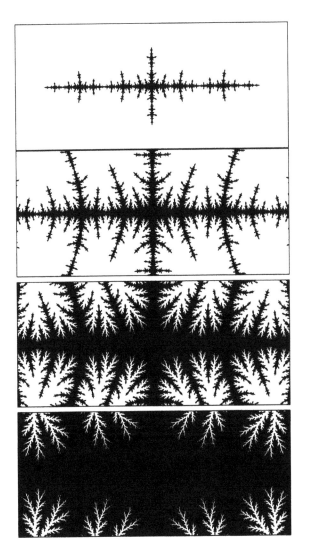

Figure 8.2. Blowups of the Feigenbaum Julia set.

Construct a fine tower $\mathcal{T} = \langle f_s : U_s \to V_s; s \in \mathcal{SR}(f) \rangle$ with $f_1 = f$, whose fine tower constants depend only on (m, B). For all $z \neq 0$ sufficiently small, there is a U_s containing z with $\text{diam}(U_s) \asymp |z|$. Then $\text{diam}\, U_s \asymp \text{diam}\, J(f_s)$, and $\log \text{diam}\, J(f_s) \asymp \log(1/s)$ by Proposition 8.1, so $\log s \asymp \log(1/|z|)$.

By Proposition 6.11 (Asymptotic density of Julia sets), there are constants D and $\eta > 0$ (depending only on (m, B)) such that

$$d_1(z, J(f_1)) < D\left(\frac{1}{s}\right)^\eta$$

where d_1 is the hyperbolic metric on $V_1 - P(f_1)$. By the Schwarz Lemma, the same bound holds in the hyperbolic metric on $\mathbb{C} - P(f)$. The hyperbolic metric $\rho(z)|dz|$ on $\mathbb{C} - P(f)$ is comparable to the metric $|dz|/d(z, P(f))$, where d denotes Euclidean distance (Corollary 8.5). Using the fact that $\log s \asymp \log(1/|z|)$ and that $d(z, P(f)) \leq d(z, 0) = |z|$, we have

$$\frac{d(z, J(f))}{|z|} = O(|z|^\alpha)$$

for some $\alpha(m, B) > 0$; equivalently $d(z, J(f)) = O(|z|^{1+\alpha})$.

Now consider $\tau \gg 0$, and let $H \subset \mathbb{H}^3$ be a supporting hyperplane for the convex hull of the Julia set, such that

$$d(\gamma(\tau), H) = d(\gamma(\tau), \partial \text{hull}(J)).$$

Since $\gamma(\tau)$ is near $0 \in \mathbb{C}$, H meets the sphere at infinity in a Euclidean circle $S(z, r)$ with $|z|$ small and $r = d(z, J(f))$. By the preceding discussion, $r = O(|z|^{1+\alpha})$. Thus H lies below the graph of $y = C|z|^{1+\alpha}$ for some constant C. But by Lemma 8.9, this means there is an $\epsilon(m, B) > 0$ such that H is disjoint from the hyperbolic ball of radius $\epsilon\tau$ about $\gamma(\tau)$.

Decreasing ϵ if necessary, the bound holds for small τ as well.

$\blacksquare$

Example. Figure 8.2 shows blowups of the Feigenbaum Julia set about its critical point by powers α^n, $n = 0, 2, 4, 6$, where $\alpha = 2.50290\ldots$ is the asymptotic self-similarity factor at the origin. (More precisely, the figures represent 1-pixel neighborhoods of the Julia set, which can be quite thick even though the Julia set is nowhere dense.)

Using generalized towers (§8.4) one can see more generally that all points in the postcritical set are deep.

8.3   Small Julia sets everywhere

In this section we show that at every scale around every point in the Julia set $J(f)$, one can find a bounded distortion copy of one of the small Julia sets $J_n(f)$. In fact, this copy is the preimage of $J_n(f)$ under some iterate f^i. The mapping f^i conjugates f^n to a quadratic-like map g defined on a definite neighborhood of $J_n(f)$.

Theorem 8.10 (Small Julia sets everywhere) *Let $f(z) = z^2 + c$ be infinitely renormalizable with combinatorics bounded by B and moduli bounded by m. Then there exist positive constants m' and λ, depending only on (m, B), such that the following holds:*

For every z in the Julia set $J(f)$, and every $r \in (0, 1]$, there is a quadratic-like map $g : U \to V$ such that

1. *g has definite modulus: $\mathrm{mod}(U, V) \geq m'$;*

2. *the Euclidean distance from z to $J(g)$ is $O(r)$: $d(z, J(g)) \leq \lambda r$;*

3. *the diameter of $J(g)$ is comparable to r:*

$$\frac{r}{\lambda} < \mathrm{diam}(J(g)) < \lambda r; \quad and$$

4. *the map g belongs to the dynamics generated by f: for some $i \geq 0$ and $n \in \mathcal{SR}(f)$, f^i maps V univalently to a neighborhood of $J_n(f)$, and conjugates g to f^n.*

Proof. All bounds in the course of the proof will depend only upon (m, B).

We begin by choosing *unbranched* renormalizations $f^n : U_n \to V_n$ such that $\mathrm{mod}(U_n, V_n) \geq m' > 0$ for all n; this is possible by Proposition 8.1.

For $1 \leq i \leq n$, the map $f^{n-i} : J_n(i) \to J_n(n) = J_n$ is injective, with a univalent inverse defined on V_n; pulling V_n and U_n back by this inverse, we obtain quadratic-like maps $f^n : U_n(i) \to V_n(i)$ (conjugate to $f^n : U_n \to V_n$ by f^{n-i}). Then $\text{mod}(U_n(i), V_n(i)) \geq m' > 0$ for all i and n.

We now construct the mapping g for various $z \in J(f)$ and $r \in (0, 1]$.

Case I: $z \in P(f)$. Each point z in the postcritical set belongs to a nested sequence of Julia sets $J(f) = J_1(1) \supset J_{n_1}(i_1) \supset J_{n_2}(i_2) \supset \cdots$ where $\mathcal{SR}(f) = \{1, n_1, n_2, \dots\}$ and the diameters of these small Julia sets tend to zero. On the other hand, $n_{i+1}/n_i \leq B$, so by Proposition 4.13, Julia sets at adjacent levels have comparable diameter. Since $\text{diam } J(f) \asymp 1$, we have small Julia sets at every scale less than one about z. Thus we can find a small Julia set $J_n(i)$ with $z \in J_n(i)$ and $\text{diam } J_n(i) \asymp r$, and take $g = f^n : U_n(i) \to V_n(i)$. This establishes the Theorem when z belongs to the postcritical set.

Case II: $z \notin P(f)$, $d(z, P(f)) < r$. This case follows immediately from Case I. Let z' be a point in $P(f)$ with $|z - z'| = d(z, P(f))$; then the g just constructed for z' also works for z, since $d(z, J(g)) \leq d(z, z') < r$.

We now treat the remaining situation: $z \notin P(f)$ and $d(z, P(f)) \geq r$. The issue here is that if r is much less than the distance of z from $P(f)$, there is no obvious quadratic-like map in the dynamics at scale r about z. To obtain g, we will study the iterates $f^k(z)$. We will show that f^k is eventually expanding enough that there is a univalent preimage of some $J_n(i)$ at scale r near z.

Let $\rho(z)|dz|$ denote the hyperbolic metric on the Riemann surface $\mathbb{C} - P(f)$. Recall $\rho(z) \asymp 1/d(z, P(f))$, where d denotes the Euclidean distance (Corollary 8.5). Let v be a tangent vector to $\mathbb{C}$ at z with Euclidean length $|v| = r$. Then the hyperbolic length $\ell(v)$ on $\mathbb{C} - P(f)$ satisfies

$$\ell(v) \asymp \frac{|v|}{d(z, P(f))} \leq 1.$$

Let $z_k = (f^k)(z)$ and let $v_k \in T_{z_k}\mathbb{C}$ be the image of v under the derivative $(f^k)'$. Consider the sequence of lengths $\ell(v_k)$; if it happens that $z_k \in P(f)$, set $\ell(v_k) = +\infty$. It is known that f is expanding in the hyperbolic metric on $\mathbb{C} - P(f)$, and that its expansion at a given point in the Julia set tends to infinity under iteration [Mc4, Thm. 3.6]. Thus $\ell(v_0) \leq \ell(v_1) \leq \ell(v_2) \dots$ and $\ell(v_k) \to \infty$. Since

$\ell(v_0) = O(1)$, there exists a $k \geq 0$ such that either $\ell(v_k) \asymp 1$, or $\ell(v_k) \ll 1$ and $\ell(v_{k+1}) \gg 1$.

Case III: the hyperbolic length jumps past 1. By this we mean $\ell(v_k) < \epsilon$ and $\ell(v_{k+1}) > 1/\epsilon$, where $\epsilon > 0$ is a small constant depending only on (m, B); its size will be determined in the course of the argument. We allow $\ell(v_{k+1}) = +\infty$.

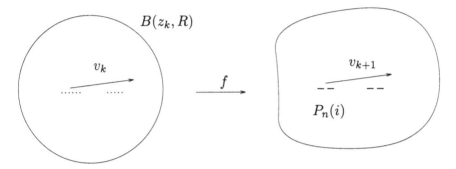

Figure 8.3. When $\ell(v_k) \ll 1$ and $\ell(v_{k+1}) \gg 1$, it is easy to pull back a copy of $J_n(i)$.

First, since $\ell(v_k) < \epsilon$, and the hyperbolic metric is comparable to the $1/d$ metric, the Euclidean distance from z_k to the postcritical set is much larger than $|v_k|$, the Euclidean length of v_k. But since the critical point $z = 0$ of $f(z) = z^2 + c$ is contained in $P(f)$, this means that f is injective on the ball $B(z_k, R)$ where $R = d(z_k, P(f)) \gg |v_k|$. Thus $f(B(z_k, R))$ contains a ball $B(z_{k+1}, R')$, where $R' \gg |v_{k+1}|$.

Now $\ell(v_{k+1})$ is large, so $|v_{k+1}| \gg d(z_{k+1}, P(f))$. By the same reasoning as in case II, there is a Julia set $J_n(i)$ whose diameter is comparable to $|v_{k+1}|$ and whose distance from z_{k+1} is $O(|v_{k+1}|)$. For ϵ small enough, this Julia set is contained in the ball $B(z_{k+1}, R'/2)$.

We now check that $J_n(i)$ can be pulled back to z_0. For $i > 1$, the sets $P_n(i)$, $U_n(i)$, $V_n(i)$ and $J_n(i)$ each have two preimages under $f(z) = z^2 + c$, namely $P_n(i-1)$, $U_n(i-1)$, $V_n(i-1)$ and $J_n(i-1)$, and the negatives of these sets, which we denote by $P'_n(i-1)$, $U'_n(i-1)$, $V'_n(i-1)$ and $J'_n(i-1)$. By our assumption that the renormalizations are unbranched, the primed sets are disjoint from the postcritical set, so all their further preimages under f are univalent.

Since $B(z_k, R)$ is disjoint from the postcritical set, we have

$$f^{-1}(J_n(i)) \cap B(z_k, R) = J'_n(i-1).$$

Since $\operatorname{diam}(J_n(i)) \asymp |v_{k+1}|$ and $d(z_{k+1}, J_n(i)) = O(|v_{k+1}|)$, the Koebe distortion theorem implies the diameter of $J'_n(i-1)$ is comparable to $|v_k|$ and its distance from z_k is $O(|v_k|)$. There is a univalent branch of f^{-k} defined on $B(z_k, R)$ and sending z_k back to $z_0 = z$; applying Koebe again, we conclude that $f^{-k}(J'_n(i-1))$ has diameter comparable to $r = |v_0|$ and its distance from z is $O(r)$ as desired. The quadratic-like map $(-f^n) : U'_n(i-1) \to V'_n(i-1)$ lifts to a quadratic-like map $g : U \to V$ with Julia set $f^{-k}(J'_n(i-1))$, completing the construction.

Case IV: the hyperbolic length $\ell(v_k) \asymp 1$. More precisely, $\ell(v_k) \in [\epsilon, 1/\epsilon]$, where ϵ is the constant in case III. Only this final case is tricky.

Suppose we can find a path δ in $\mathbb{C} - P(f)$ from z_k to $\gamma_n(i)$ (for some n, i) such that $\ell(\delta)$ is bounded, and such that the germ h of f^{-k} with $h(z_k) = z_0$ can be analytically continued along δ to a single-valued function on $\gamma_n(i)$. Then h extends across $P_n(i)$, and then from $P_n(i)$ to $V_n(i)$.

Let $g : U \to V$ be obtained by conjugating $f^n : U_n(i) \to V_n(i)$ by h. We claim g satisfies the Theorem for $z = z_0$. Indeed, $\operatorname{diam} J_n(i)$ and $|v_k|$ are comparable, because $\operatorname{diam} J_n(i) \asymp \operatorname{diam} P_n(i)$ by Proposition 8.1, $\operatorname{diam} P_n(i) \asymp \operatorname{diam}(\gamma_n(i))$ because $\ell(\gamma_n(i)) \asymp 1$, and $|v_k| \asymp \operatorname{diam}(\gamma_n(i))$ because $\ell(v_k) \asymp 1$ and the hyperbolic distance from z_k to $\gamma_n(i)$ is bounded. Similarly, the Euclidean distance from z_k to $J_n(i)$ is $O(|v_k|)$.

Using the lower bound on the injectivity radius, we can apply Koebe to a bounded number of balls covering $\delta \cup \gamma_n(i)$ to conclude that $|h'| \asymp |h'(z_k)| = |v_0|/|v_k|$ along $\gamma_n(i)$. Applying Koebe twice more, we find $|h'| \asymp |v_0|/|v_k|$ on $P_n(i)$, and then on $J_n(i)$. Thus $\operatorname{diam} J(g) = \operatorname{diam} h(J_n(i)) \asymp |v_0| = r$ and $d(z_0, J(g)) = O(r)$ as stated in the Theorem.

It remains only to construct δ. Consider the decomposition of the Riemann surface $\mathbb{C} - P(f)$ into planar pieces with bounded geometry obtained by slicing along the curves $\Gamma = \{\gamma_n(i) : n \in \mathcal{SR}(f), n > 1, 1 \le i \le n\}$ (as in Figure 8.1). One such piece contains z_k. Pick

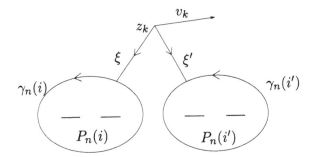

Figure 8.4. When $\ell(v_k) \asymp 1$, we can pull back a copy of either $J_n(i)$ or $J_n(i')$.

two "lower" boundary components $\gamma_n(i)$ and $\gamma_n(i')$ of this piece, and join them to z_k by geodesic segments ξ, ξ' of minimal length (see Figure 8.4). We will assume $i' > i$.

By the bounded geometry of $\mathbb{C} - P(f)$, the hyperbolic lengths of ξ and ξ' are bounded (in terms of (m, B)). If f^{-k} has a univalent branch along $\xi \cup \gamma_n(i)$, sending z_k to z_0, then we may take $\delta = \xi$ and the proof is complete. Similarly the proof is complete if f^{-k} has a univalent branch along $\xi' \cup \gamma_n(i')$.

If both ξ and ξ' fail, we claim the path $\delta = \xi' * \xi^{-1} * \gamma_n(i) * \xi$ succeeds. This path joins z_k to $\gamma_n(i')$ after running once around $\gamma_n(i)$.

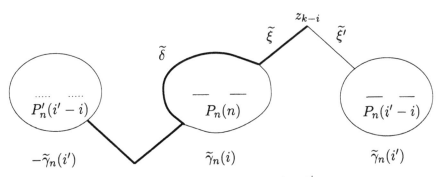

Figure 8.5. Preimages under f^i.

To see that δ works, consider the greatest integer $j < k$ such that f^{-j} has a univalent branch on $\xi \cup \gamma_n(i)$, sending z_k back to z_{k-j}. For this branch, $\alpha = f^{-j}(\gamma_n(i))$ encloses $P_n(1)$, the part of the postcritical set containing the critical value of f — otherwise both branches of f^{-1} would be univalent on α. Consequently $j = i - 1$. Since $i' > i$, the loop $\beta = f^{-j}(\gamma_n(i'))$ is a univalent preimage of $\gamma_n(i')$. Thus β encloses $P_n(i' - j)$; otherwise it would bound a disk disjoint from the postcritical set, on which all inverse branches of f are univalent.

Now let $\tilde{\xi}$, $\tilde{\xi}'$ and $\tilde{\delta}$ be the images of ξ, ξ' and δ under the branch of f^{-i} sending z_k to z_{k-i}. Then $\tilde{\xi}$ joins z_{k-i} to $\tilde{\gamma}_n(i) = f^{-1}(\alpha)$, the latter loop covering α by degree two and enclosing the critical point of f. The path $\tilde{\xi}'$ terminates at the component $\tilde{\gamma}_n(i')$ of $f^{-1}(\beta)$ enclosing $P_n(i' - i)$. But since δ wraps once around $\gamma_n(i)$, its lift $\tilde{\delta}$ starting at z_{k-i} terminates at the *other* preimage $-\tilde{\gamma}_n(i')$ of β, which bounds a disk disjoint from the postcritical set (see Figure 8.5.) Thus all branches of f^{i-k} are univalent on $\tilde{\delta} \cup -\tilde{\gamma}_n(i')$, so f^{-k} is univalent on $\gamma_n(i') * \delta$, completing the proof.

∎

Remarks. The argument above is closely related to the proof of the rigidity of robust maps $f(z)$ given in [Mc4, Thm. 10.10]. To establish quasiconformal rigidity, it suffices to construct quadratic-like dynamics about every $z \in J(f)$ and at infinitely many distinct scales. The proof above uses bounded combinatorics to obtain quadratic-like dynamics at *every* scale.

With this result in hand, we can easily reprove the rigidity of f and the rigidity of towers; we restate these results as corollaries.

Corollary 8.11 *The map f admits no invariant line field on its Julia set.*

Proof. If the Julia set of f admits an invariant line field μ, we can choose a point $z \in J(f)$ and a scale $r > 0$ such that the line field is nearly parallel in a ball of radius r about z. By the result above, there is a quadratic-like map $g : U \rightarrow V$ defined near z at scale r with $\text{mod}(U, V) \geq m' > 0$, such that $g^*(\mu) = \mu$. By compactness of

$Poly_2(m')$, we can rescale and pass to a limit to obtain a quadratic-like map which leaves invariant a family of parallel lines in the plane. This is clearly impossible.

■

Corollary 8.12 *A bi-infinite tower $\mathcal{T}$ with bounded combinatorics and definite moduli is quasiconformally rigid.*

Proof. Given an invariant line field for $\mathcal{T}$, choose a small disk D in which the line field is nearly parallel. The union of the Julia sets $J(f_s)$ for mappings f_s in $\mathcal{T}$ is dense in $\mathbb{C}$, so $J(f_s)$ comes close to the center of D for some s; taking s small enough, we can assume $\operatorname{diam} J(f_s) > \operatorname{diam} D$ as well. The Theorem above yields a quadratic-like map $g : U \to V$ with $J(g) \subset D$ and $\operatorname{diam} J(g) \asymp \operatorname{diam} D$, under which μ must also be invariant, a contradiction.

■

8.4   Generalized towers

The present theory of towers is intended to capture geometric limits of the iterates of a single quadratic mapping f under rescaling about its critical point. A similar theory exists for arbitrary points in the postcritical set $P(f)$. Here we sketch the results that follow from a more general development.

A *generalized tower* is a collection $\mathcal{T} = \langle f_s; \ s \in S \rangle$ of maps $f_s \in Poly_2$, indexed by a level set S, such that:

> *whenever $n = t/s > 1$, f_s^n is simply renormalizable, $K(f_t) = K_n(i, f_s)$ for some i, and $f_s^n = f_t$ on $K(f_t)$.*

The basic example comes from an infinitely renormalizable map $f \in Poly_2$ with a specified point p in its postcritical set $P(f)$. A generalized tower can be used to record the dynamics on the small Julia sets nesting around p. To construct it, set $S = \mathcal{SR}(f)$, and for each level $s \in S$, choose a renormalization $f^s : U_s \to V_s$; then let $f_s = f^s : U_s(i) \to V_s(i)$, where $p \in J_s(i, f)$.

The towers we have used before correspond to the case where p is the critical point of f_1. The arguments already presented can be adapted to prove:

Theorem 8.13 *Let T be a bi-infinite generalized tower with bounded geometry and definite moduli. Then:*

1. *T is quasiconformally rigid.*

2. *The union of the Julia sets of T is dense in $\mathbb{C}$.*

3. *T is determined up to isomorphism by its inner classes $I(f_s)$ and by the integers $i(s,t)$ such that $J(f_t) = J_{t/s}(i(s,t), f_s)$.*

4. *For $n > 1$ and $1 \leq i \leq n$, define a generalized renormalization operator $\mathcal{R}_{n,i} : \mathcal{G}^{(n)} \to \mathcal{G}$ by mapping $[f]$ to $[f^n : U_n(i) \to V_n(i)]$. Suppose $\mathcal{R}_n$ has a fixed point F. Then $\mathcal{R}_{n,i}$ has a unique fixed point F_i with inner class $I(F_i) = I(F)$, and $\mathcal{R}_{n,i}^k(g) \to [F_i]$ for all germs g with the same inner class as F.*

   The germs F and F_i are conformally conjugate.

5. *Let $f(z) = z^2 + c$ be infinitely renormalizable, with bounded combinatorics and definite moduli. Then every $x \in P(f)$ is a deep point of $J(f)$.*

Remarks. The conformal conjugacy between the renormalization fixed points F and F_i results from the fact that f^{n-i} provides a conformal conjugacy between $f^n : U_n \to V_n$ and $f^n : U_n(i) \to V_n(i)$, for any renormalization of a quadratic polynomial [Mc4, Thm. 7.2].

   The generalized renormalization operators $\mathcal{R}_{n,i}$ and their fixed points seem to have only rarely appeared in the literature. See however [CoTr] for a discussion of renormalization around the critical value, rather than the critical point, in the case of period doubling (this corresponds to the operator $\mathcal{R}_{2,1}$). This reference was provided by H. Epstein.

9 Geometric limits in dynamics

In this chapter we formulate a general theory of holomorphic dynamical systems. The setting will be broad enough to include both Kleinian groups and iterated rational maps. Its scope will also encompass geometric limits of rational maps; as we have seen, renormalization naturally leads from polynomials to polynomial-like maps, and it is useful to allow quite general transformations in the limiting dynamics.

We then prove inflexibility theorems for dynamical systems which generalize our earlier results for Kleinian groups. The injectivity bounds in Theorems 2.11 and 2.18 are replaced by the assumption of *uniform twisting*, formulated in §9.3 below.

As an application, we sharpen Theorem 7.9 by showing the renormalizations $\mathcal{R}_p^n(f)$ converge *exponentially fast* to the renormalization fixed point F. In particular, this fast convergence holds for the Feigenbaum polynomial f. Exponential convergence implies that f is $C^{1+\alpha}$ conjugate to F on its postcritical set, one of the main renormalization conjectures.

9.1 Holomorphic relations

Definitions. We adopt the convention that a *complex manifold* has at most countably many components, all of the same dimension.

Let Z be a complex manifold. A set $F \subset Z$ is an *analytic hypersurface* if for each $p \in F$ there is a neighborhood U of p in Z and a nonconstant holomorphic map $f : U \to \mathbb{C}$ such that $F \cap U = f^{-1}(0)$. We do *not* require F to be closed in Z.

Now let X be a complex 1-manifold. A *holomorphic relation* $F \subset X \times X$ is a set which can be expressed as a countable union of analytic hypersurfaces. A (one-dimensional) *holomorphic dynamical system* $(X, \mathcal{F})$ is a collection $\mathcal{F}$ of holomorphic relations on a complex 1-manifold X.

Holomorphic relations are composed by the rule

$$F \circ G = \{(x,y) \; : \; (x,z) \in G \text{ and } (z,y) \in F \text{ for some } z \in X\},$$

and thereby give rise to dynamics.

For any holomorphic relation F, there is a complex 1-manifold $\widehat{F}$ and a holomorphic map

$$\nu : \widehat{F} \to X \times X$$

such that $\nu(\widehat{F}) = F$ and ν is injective outside a countable subset of $\widehat{F}$. The surface $\widehat{F}$ is the *normalization* of F; it is unique up to isomorphism over F [Gun2].

Examples. A holomorphic map $f : U \to X$, $U \subset X$ gives a holomorphic relation by identifying f with its graph

$$\mathrm{gr}(f) = \{(x, f(x)) \ : \ x \in U\} \subset X \times X.$$

The inverse of f (obtained by interchanging factors) is also a holomorphic relation (generally multivalued), as is

$$\mathrm{gr}(f^{-i} \circ f^j) = \{(z, w) \ : \ f^i(w) = f^j(z)\}.$$

A Kleinian group, the iterates of a rational map, a polynomial-like map and the collection of maps appearing in a tower are all examples of holomorphic dynamical systems.

Conformal and quasiconformal conjugacy. Let $\alpha : X \to Y$ be a biholomorphic map between complex 1-manifolds. Given a holomorphic dynamical system $\mathcal{F}$ on X,

$$\mathcal{G} = \alpha_*(\mathcal{F}) = \{(\alpha \times \alpha)(F) \ : \ F \in \mathcal{F}\}$$

is a holomorphic dynamical system on Y. The map α provides a *conformal conjugacy* between $(X, \mathcal{F})$ and $(Y, \mathcal{G})$.

Let $M(X)$ denote the Banach space of L^∞ Beltrami differentials on X with $\|\mu\| = \mathrm{ess.\,sup}\,|\mu|$. Let $\nu : \widehat{F} \to F$ be the normalization of a holomorphic relation F, and let $\pi_i : X \times X \to X$ denote projection to the ith factor. Then $\mu \in M(X)$ is F-*invariant* if

$$(\pi_1 \circ \nu)^* \mu = (\pi_2 \circ \nu)^* \mu$$

on every component of $\widehat{F}$ where both $\pi_1 \circ \nu$ and $\pi_2 \circ \nu$ are nonconstant. Equivalently, $h^*(\mu) = \mu|U$ for every univalent map $h : U \to V$ such that the graph of h is contained in F. (We regard two differentials agreeing outside a set of measure zero as equal.)

The Beltrami differential μ is $\mathcal{F}$-*invariant* if it is invariant under every relation F in $\mathcal{F}$. If $\|\mu\|_\infty < 1$, there is a Riemann surface Y and a quasiconformal map $\phi : X \to Y$ with complex dilatation μ; and μ is $\mathcal{F}$-invariant if and only if

$$\mathcal{G} = \phi_*(\mathcal{F}) = \{(\phi \times \phi)(F) \ : \ F \in \mathcal{F}\}$$

is a holomorphic dynamical system on Y. If so, ϕ is a *quasiconformal conjugacy* from $(X, \mathcal{F})$ to $(Y, \mathcal{G})$.

The space of hypersurfaces. To describe a topology on the space of holomorphic dynamical systems, we first put a topology on the space of relations.

Let Z be a complex manifold, and let $\mathcal{V}(Z)$ denote the set of all analytic hypersurfaces in Z. Thus $\mathcal{V}(\widehat{\mathbb{C}} \times \widehat{\mathbb{C}})$ includes the graphs of all Möbius transformations, rational maps, algebraic correspondences, entire functions, etc., but excludes holomorphic relations such as

$$\mathrm{gr}(z^{\sqrt{2}}) = \{(e^t, e^{\sqrt{2}t}) \ : \ t \in \mathbb{C}\}.$$

We topologize $\mathcal{V}(Z)$ as follows. First, given $F \in \mathcal{V}(Z)$, let $\partial F = \overline{F} - F$. Then ∂F is closed and F is a closed hypersurface in $Z - \partial F$. We say $F_i \to F$ in $\mathcal{V}(Z)$ if

(a) $\partial F_i \to \partial F$ in the Hausdorff topology on closed subsets of Z, and

(b) for any $z \in Z - \partial F$, there is a neighborhood U of z and a collection of holomorphic functions $f, f_i : U \to \mathbb{C}$ such that

$$F \cap U = f^{-1}(0), \quad F_i \cap U = f_i^{-1}(0) \quad \text{for all } i \gg 0,$$

f vanishes to order one on $F \cap U$, and $f_i \to f$ uniformly on U.

Example: the closure of the group of Möbius transformations. Let $\mathrm{Aut}(\widehat{\mathbb{C}}) \subset \mathcal{V}(\widehat{\mathbb{C}} \times \widehat{\mathbb{C}})$ be the inclusion which sends a Möbius transformation to its graph. Then

$$\overline{\mathrm{Aut}(\widehat{\mathbb{C}})} = \mathrm{Aut}(\widehat{\mathbb{C}}) \cup \bigcup_{a,b \in \widehat{\mathbb{C}}, a \neq b} \{(w, z) \ : \ w = a \text{ or } z = b\}.$$

For example, the sequence of Möbius transformations $w = M_n(z) = 1/(nz)$ goes to infinity in $\mathrm{Aut}(\widehat{\mathbb{C}})$, but their graphs converge to the hypersurface $F = \{(z, w) \; : \; z = 0 \text{ or } w = 0\}$. Note that each point in $\overline{\mathrm{Aut}(\widehat{\mathbb{C}})}$ represents the $(1, 1)$ class in $H_2(\widehat{\mathbb{C}} \times \widehat{\mathbb{C}}, \mathbb{Z}) \cong \mathbb{Z} \oplus \mathbb{Z}$.

In passing from Kleinian groups to more general dynamical systems on the sphere, we will replace $\mathrm{Aut}(\widehat{\mathbb{C}})$ with $\mathcal{V}(\widehat{\mathbb{C}} \times \widehat{\mathbb{C}})$, since the latter includes all the relations we wish to consider. There is an important technical difference between these spaces: the space of hypersurfaces is not locally compact. For example, $\mathbb{R}_{\geq 0}^\infty$ embeds in $\mathcal{V}(\widehat{\mathbb{C}} \times \widehat{\mathbb{C}})$ with closed image by the map which sends $(t_1, t_2, \dots)$ to the hypersurface $F = (\bigcup_k B(2k, e^{-t_k})) \times \{0\}$. However, we have:

Proposition 9.1 *For any complex manifold Z, the space of analytic hypersurfaces $\mathcal{V}(Z)$ is a separable metrizable space.*

Proof. In brief, any $F \in \mathcal{V}(Z)$ defines a positive $(1, 1)$-current on any compact set disjoint from ∂F. We have $F_i \to F$ in $\mathcal{V}(Z)$ if and only if (a) $\partial F_i \to \partial F$ and (b) $F_i \to F$ as currents on each compact subset of $Z - \partial F$. Since the space of closed subsets of Z and the space of currents on $Z - \partial F$ are both separable metric spaces, so is $\mathcal{V}(Z)$.

Here is a more detailed argument. Let Z be a complex n-manifold, and let (U_k, ϕ_k) be a countable collection of open sets $U_k \subset Z$ and smooth $(n - 1, n - 1)$-forms ϕ_k on Z, such that $\overline{U_k}$ is compact, the U_k form a base for the topology on Z, ϕ_k is supported in U_k and the span of the ϕ_k is dense in the space of C^∞ compactly supported $(n - 1, n - 1)$-forms on Z. Choose a compact set $K_k \subset U_k$ whose interior contains the support of ϕ_k, and let $\rho_k : \mathrm{Cl}(Z) \to [0, 1]$ be a continuous bump function such that $\rho_k(E) = 1$ if $E \cap U_k = \emptyset$, and $\rho_k(E) = 0$ if $E \cap K_k \neq \emptyset$. (Since $\mathrm{Cl}(Z)$ is a compact metrizable space, such a function exists by Urysohn's Lemma.) Define $\alpha_k : \mathcal{V}(Z) \to \mathbb{R}$ by

$$\alpha_k(F) = \begin{cases} \rho_k(\partial F) \int_{F^*} \phi_k & \text{if } \partial F \cap \mathrm{supp}(\phi_k) = \emptyset, \\ 0 & \text{otherwise.} \end{cases}$$

Here F^* is the complex $(n - 1)$-manifold of smooth points of F.

We claim $\alpha_k : \mathcal{V}(Z) \to \mathbb{R}$ is continuous. Indeed, if $F_i \to F$, and ∂F meets the support of ϕ_k, then ∂F_i meets K_k for all $i \gg 0$, so

$\lim \alpha_k(F_i) = \alpha_k(F) = 0$. On the other hand, if $\partial F \cap \text{supp}(\phi_k) = \emptyset$, then near each $p \in \text{supp}(\phi_k)$ there are local defining functions f_i for F_i converging uniformly to a defining function f for F. Then $\log|f_i| \to \log|f|$ locally in L^1 (as can be seen using the Weierstrass preparation theorem). Therefore $\overline{\partial}\partial \log|f_i| \to \overline{\partial}\partial \log|f|$ as distributional $(1,1)$ forms; but these forms represent the currents of integration over F_i and F respectively, so $\int_{F_i^*} \phi_k \to \int_{F^*} \phi_k$. This establishes continuity of α_k.

Conversely, suppose $\partial F_i \to \partial F$ and $\alpha_k(F_i) \to \alpha_k(F)$ for all k. Then $F_i \to F$ as currents on any compact set disjoint from ∂F. Let p be a point of $Z - \partial F$ and let $\overline{\Delta^n} \subset Z - \partial F$ be a polydisk chart with $p = 0$, in which F is the zero set of a Weierstrass polynomial

$$w(z) = z_1^d + a_1(z_2, \ldots, z_n)z_1^{d-1} + \ldots + a_d(z_2, \ldots, z_n).$$

The support of F_i converges to the support of F, and the intersection number $D \cdot F = \lim D \cdot F_i$ for any disk transverse to F. It follows that F_i is represented by a Weierstrass polynomial w_i for all i sufficiently large, and for each disk $D = \Delta \times (z_2, \ldots, z_n)$, the set $D \cap F_i$ is close to $D \cap F$. Thus $w_i \to w$ uniformly on $\overline{\Delta^n}$, and $F_i \to F$ in $\mathcal{V}(Z)$.

Summarizing, the map $F \mapsto (\partial F, \alpha_k(F))$ gives an embedding $\mathcal{V}(Z) \to \text{Cl}(Z) \times \mathbb{R}^\infty$, so $\mathcal{V}(Z)$ is a separable metric space. ∎

Details on currents and the Weierstrass theory can be found in [GH]. One may also use the coefficients of Weierstrass polynomials directly to construct local separating functions on $\mathcal{V}(Z)$.

The geometric topology. In the sequel we will consider exclusively dynamical systems $(X, \mathcal{F})$ where $\mathcal{F} \subset \mathcal{V}(X \times X)$. In other words, each relation $F \in \mathcal{F}$ will be a single hypersurface (rather than a countable union of hypersurfaces). A dynamical system $\mathcal{F}$ is *closed* if it is a closed subset of $\mathcal{V}(X \times X)$.

The *geometric topology* on the space of closed dynamical systems is defined to be the Hausdorff topology on closed subsets of $\mathcal{V}(X \times X)$ (see §2.1). That is, $\mathcal{F}_i \to \mathcal{F}$ geometrically if and only if

(a) every $F \in \mathcal{F}$ is the limit of a sequence $F_i \in \mathcal{F}_i$, defined for all i sufficiently large; and

(b) if $F = \lim F_{i_k}$ for a subsequence $F_{i_k} \in \mathcal{F}_{i_k}$, then $F \in \mathcal{F}$.

By definition, a collection of dynamical systems

$$\{\mathcal{F}_\alpha\} \subset \mathrm{Cl}(\mathcal{V}(X \times X))$$

is closed in the geometric topology if it contains the limit of every convergent sequence $\mathcal{F}_{\alpha(i)}$.

Some care is warranted when dealing with this topology, since $\mathcal{V}(X \times X)$ is not locally compact. In general, when E is not locally compact, $\mathrm{Cl}(E)$ is not Hausdorff. Also, the closure of a set is not always obtained by adjoining limits of all sequences in the set; in general the limits must be iterated. For example, in $\mathrm{Cl}(\mathbb{N}^\mathbb{N})$, let

$$F_{ij} = \{a_n \ : \ a_1 \geq i \text{ or } a_2 \geq j \text{ or } a_3 \geq j \text{ or } \ldots \ a_i \geq j\}.$$

Then

$$\lim_{j\to\infty} F_{ij} = F_i = \{a_n \ : \ a_1 \geq i\}$$

and $\lim_{i\to\infty} F_i = \emptyset$, but $\lim F_{i(k)j(k)} \neq \emptyset$ for any convergent subsequence of F_{ij}.

On the other hand, the Hausdorff topology on the closed subsets of a separable metric space is sequentially compact (Proposition 2.1), so we have:

Corollary 9.2 *Every sequence $\mathcal{F}_i$ of closed dynamical systems has a geometrically convergent subsequence.*

Proposition 9.3 *If μ_i is F_i-invariant, $\mu_i \to \mu$ in the weak* topology on $M(X)$, and $F_i \to F$ in $\mathcal{V}(Z)$, then μ is F-invariant.*

Proof. It suffices to show $h^*(\mu) = \mu|U$ for all univalent maps $h :$ $U \to X$ such that U is a disk in X and the graph of h is contained in a compact subset of the smooth points of F. Near any point of the graph of h, F is defined by a equation of the form $f(z_1, z_2) = z_2 - h(z_1)$ in local product coordinates. By the implicit function theorem, for all i sufficiently large, F_i contains the graph of a function $h_i : U \to X$ such that $h_i \to h$ uniformly on U. Then $h_i^*(\mu_i) = \mu_i|U$, so in the limit $h^*(\mu) = \mu|U$. ∎

Corollary 9.4 *If $\mathcal{F}_i \to \mathcal{F}$ in the geometric topology, $\mu_i \to \mu$ in the weak* topology on $M(X)$, and μ_i is $\mathcal{F}_i$-invariant, then μ is $\mathcal{F}$-invariant.*

Saturation. Given a set $\mathcal{F} \subset \mathcal{V}(X \times X)$, we define its *saturation* by

$$\mathcal{F}^{\text{sat}} = \overline{\{F \cap U \ : \ F \in \mathcal{F} \text{ and } U \text{ is an open subset of } X \times X\}}.$$

If $\mathcal{F} = \mathcal{F}^{\text{sat}}$, we say $\mathcal{F}$ is *saturated*. Saturation is preserved under geometric limits. If μ is $\mathcal{F}$-invariant, it is also $\mathcal{F}^{\text{sat}}$-invariant.

We introduce the saturation to take account of convergence on subdomains in X, even when the original transformations are defined on all of X.

Proposition 9.5 (Saturated convergence) *Let $\mathcal{F}_i \to \mathcal{F}$ be a geometrically convergent sequence of saturated dynamical systems. Then the graph of $f : U \to X$ belongs to $\mathcal{F}$ if and only if there is a sequence $f_i : U_i \to X$ with $\mathrm{gr}(f_i) \in \mathcal{F}_i$ for all i sufficiently large, such that $f_i \to f$ uniformly on compact subsets of U.*

(In particular any compact subset of U is eventually contained in U_i.)

Proof. Let V_k be a increasing sequence of domains with $\overline{V_k} \subset U$ compact, such that $U = \bigcup V_k$. Then $\mathrm{gr}(f|V_k) \to \mathrm{gr}(f)$ in $\mathcal{V}(X \times X)$.

If $\mathrm{gr}(f) \in \mathcal{F}$, then by saturation $\mathrm{gr}(f|V_k) \in \mathcal{F}$ for every k. By the implicit function theorem, there is a sequence of mappings $f_{i,k} : V_k \to X$ with $\mathrm{gr}(f_{i,k}) \in \mathcal{F}_i$ for all $i \gg 0$, and with $f_{i,k} \to f|V_k$ uniformly as $i \to \infty$. Diagonalizing, we obtain a sequence $f_{i,k(i)} : V_{k(i)} \to X$, such that $f_{i,k(i)} \to f$ uniformly on compact subsets of U. Setting $U_i = V_{k(i)}$ and $f_i = f_{i,k(i)}$ completes the proof in one direction.

Conversely, given $f_i : U_i \to X$ converging to f on compact sets, it is evident that $\mathrm{gr}(f_i|V_k) \to \mathrm{gr}(f|V_k)$ and thus $\mathrm{gr}(f|V_k) \in \mathcal{F}$. But $\mathcal{F}$ is closed, so $\mathrm{gr}(f) \in \mathcal{F}$ as well.

∎

Example: Kleinian groups. We now have two notions of geometric convergence for Kleinian groups. Let Γ_n be a sequence of Kleinian groups, and suppose $\Gamma_n^{\text{sat}} \to \mathcal{F}$ as closed subsets of $\mathcal{V}(\widehat{\mathbb{C}} \times \widehat{\mathbb{C}})$. Then it is easy to see that $\Gamma_n \to \Gamma = \mathcal{F} \cap \mathrm{Aut}(\widehat{\mathbb{C}})$ in the traditional sense of geometric convergence (coming from the Hausdorff topology on $\mathrm{Cl}(\mathrm{Aut}(\widehat{\mathbb{C}}))$; see §2.2).

On the other hand, $\mathcal{F}$ generally contains more information than the traditional geometric limit Γ, since it records limits of Möbius transformations tending to infinity in $\mathrm{Aut}(\widehat{\mathbb{C}})$. For example, the groups Γ_n generated by $\gamma_n(z) = nz$ converge to the trivial group in the traditional sense, but the relation $\{(z,w) : w = \infty \text{ or } z = 0\}$ is included in the geometric limit $\mathcal{F}$.

9.2 Nonlinearity and rigidity

In this section we consider holomorphic dynamical systems $\mathcal{F}$ on the Riemann sphere $\widehat{\mathbb{C}}$. The group $\mathrm{Aut}(\widehat{\mathbb{C}})$ of Möbius transformations acts on these dynamical systems, and in some sense the full orbit $\mathrm{Aut}(\widehat{\mathbb{C}}) \cdot \mathcal{F}$ is a replacement for the hyperbolic 3-manifold $M = \mathbb{H}^3/\Gamma$ of a Kleinian group.

By combining this idea with that of geometric limits, we will say what it means for $\mathcal{F}$ to be *uniformly nonlinear* on a set $\Lambda \subset \widehat{\mathbb{C}}$. This notion generalizes upper and lower bounds on the injectivity radius of a hyperbolic manifold in its convex core. It is then straightforward to prove an infinitesimal inflexibility theorem for dynamical systems.

In the next section we integrate these bounds and obtain inflexibility of conjugacies.

Nonlinearity. Let $\mathcal{F} \subset \mathcal{V}(\widehat{\mathbb{C}} \times \widehat{\mathbb{C}})$ be a holomorphic dynamical system on the Riemann sphere. We say $\mathcal{F}$ is *linear* if it leaves invariant a parabolic line field; otherwise, $\mathcal{F}$ is *nonlinear*. If $\mathcal{F}$ is linear, then after changing coordinates by a Möbius transformation, it preserves the line field $\mu = d\bar{z}/dz$. In these coordinates $h'(z)$ is constant and real for any univalent map $h : U \to \widehat{\mathbb{C}}$ with $\mathrm{gr}(h) \subset F \in \mathcal{F}$.

It is convenient to have a quantitative measure for the nonlinearity of a dynamical system. To define one, let

$$\sigma(z)|dz| = \frac{2|dz|}{1 + |z|^2}$$

be the spherical metric, and let $B(x, r) \subset \widehat{\mathbb{C}}$ denote the spherical ball of radius r centered at x. For $0 < r < 1$, let $\mathcal{U}_r$ be the set of all univalent maps $f : (B(x, r), x) \to \widehat{\mathbb{C}}$ such that $r \leq \|f'(x)\|_\sigma \leq 1/r$, and $B(f(x), r^2) \subset f(B(x, r))$. Using the Koebe distortion theorem, it is easy to see that $\mathcal{U}_r$ is compact in the Carathéodory topology. Define $\nu(\mathcal{F})$, the *nonlinearity* of $\mathcal{F}$, by

$$\nu(\mathcal{F}) = \inf_{\{\text{parabolic } \mu\}} \quad \sup_{\substack{\{f \in \mathcal{U}_r, 0 < r < 1, \\ \text{with } \mathrm{gr}(f) \in \mathcal{F}^{\text{sat}}\}}} \int_{B(x, r)} |\mu - f^* \mu| \, \sigma^2(z) |dz|^2.$$

In other words, $\nu(\mathcal{F}) \geq \delta$ if and only if, for any parabolic line field μ, there is an $f : B(x, r) \to \widehat{\mathbb{C}}$ in $\mathcal{U}_r$ with the graph of f in the saturation of $\mathcal{F}$, such that the L^1-deviation of μ from $f^* \mu$ (with respect to spherical area on $B(x, r)$) is at least δ.

Proposition 9.6 *The nonlinearity $\nu(\mathcal{F}) = 0$ if and only if $\mathcal{F}$ is linear. If $\mathcal{F}_n \to \mathcal{F}$ is a geometrically convergent sequence of saturated dynamical systems, then $\nu(\mathcal{F}_n) \to \nu(\mathcal{F})$.*

Proof. These assertions follow easily from compactness of the space of parabolic line fields, compactness of $\mathcal{U}_r$ and Proposition 9.5.

■

Obviously $\nu(\mathcal{F}) = \nu(\mathcal{F}^{\text{sat}})$, so saturation is unimportant when measuring the nonlinearity of a fixed dynamical system. On the other hand, ν is not continuous under geometric limits without the assumption of saturation. For example, $\mathcal{F}_n = \{\mathrm{gr}(z^2 + z^n)\}$ converges geometrically to $\mathcal{F} = \emptyset$, a linear dynamical system, even though $\liminf \nu(\mathcal{F}_n) > 0$.

Uniform nonlinearity. A collection of dynamical systems $\{\mathcal{F}_\alpha\}$ on the sphere is *uniformly nonlinear* if $\lim \mathcal{F}_{\alpha(i)}^{\text{sat}}$ is nonlinear for any geometrically convergent sequence in the family of saturations.

Let

$$\nu(\{\mathcal{F}_\alpha\}) = \inf_\alpha \nu(\mathcal{F}_\alpha).$$

By continuity of the nonlinearity under geometric limits, we have:

Proposition 9.7 *A collection $\{\mathcal{F}_\alpha\}$ is uniformly nonlinear if and only if $\nu(\{\mathcal{F}_\alpha\}) > 0$.*

Corollary 9.8 *If $\{\mathcal{F}_\alpha\}$ is uniformly nonlinear, so is $\overline{\{\mathcal{F}_\alpha^{\mathrm{sat}}\}}$, where the closure is taken in the geometric topology.*

Note that the Corollary is *not* immediate from the definition of uniform nonlinearity, because in general $\{F_\alpha^{\mathrm{sat}}\}$ union its sequential limits is a proper subset of $\overline{\{\mathcal{F}_\alpha^{\mathrm{sat}}\}}$.

Families of dynamical systems $(\mathcal{F}, \Lambda)$. Let ω_0 denote the standard frame at the origin in the ball model for hyperbolic space $\mathbb{H}^3$. For any other ω in the frame bundle of $\mathbb{H}^3$, there is a unique Möbius transformation g sending ω_0 to ω. Mimicking the correspondence between manifolds with baseframe and Kleinian groups, we let $(\mathcal{F}, \omega)$ denote the dynamical system $g^*(\mathcal{F})$. The notation is meant to suggest that $(\mathcal{F}, \omega)$ is $\mathcal{F}$ as seen from ω.

For any compact set $\Lambda \subset \widehat{\mathbb{C}}$, let

$$(\mathcal{F}, \Lambda) = \{(\mathcal{F}, \omega) \; : \; \omega \text{ is a frame in the convex hull of } \Lambda\}.$$

The family of dynamical systems $(\mathcal{F}, \widehat{\mathbb{C}})$ represents the full orbit of $\mathcal{F}$ under the automorphism group of $\widehat{\mathbb{C}}$. If Γ is a Kleinian group with normalizer Γ' in $G = \mathrm{Isom}^+(\mathbb{H}^3)$, then $(\Gamma, \widehat{\mathbb{C}})$ can be naturally identified with G/Γ', the frame bundle of $M = \mathbb{H}^3/\Gamma'$. Thus $(\mathcal{F}, \Lambda)$ plays a role similar to M for more general dynamical systems.

Example: uniformly nonlinear groups. Let us define

$$\nu(\mathcal{F}, \Lambda) = \inf_{\text{frames } \omega \text{ in hull}(\Lambda)} \nu(\mathcal{F}, \omega).$$

We will show that this measure of uniform nonlinearity is like an upper and lower bound on the injectivity radius for a Kleinian group.

Recall that a Kleinian group is *elementary* if it contains an abelian subgroup of finite index.

Proposition 9.9 *Let $M = \mathbb{H}^3/\Gamma$ be a nonelementary hyperbolic manifold, and let Λ be the limit set of Γ. Then (Γ, Λ) is uniformly nonlinear if and only if the injectivity radius of M is bounded above and below in its convex core.*

Moreover

$$\nu(\Gamma, \Lambda) \; \geq \; \nu_0(R_0, R_1) > 0$$

when the injectivity radius ranges in $[R_0, R_1]$ and $R_0 > 0$.

Proof. First observe that if Γ_n is a sequence of Kleinian groups, then by passing to a subsequence we can assume $\Gamma_n \to \Gamma$ in the traditional geometric topology, and $\Gamma_n^{\mathrm{sat}} \to \mathcal{F}$ as closed subsets of $\mathcal{V}(\widehat{\mathbb{C}} \times \widehat{\mathbb{C}})$. The limit $\mathcal{F}$ is nonlinear if and only if Γ is nonlinear. (Indeed, every $F \in \mathcal{F}$ is either the restriction of an element of Γ, or a relation of the form $\{(z, w) \; : \; z = a \text{ or } w = b\}$. By definition, the latter relation preserves any Beltrami differential on the sphere.) Thus (Γ, Λ) is uniformly nonlinear if and only if every traditional geometric limit $\Gamma_\infty = \lim (\Gamma, \omega_n)$ is nonlinear.

If the injectivity radius in the convex core is unbounded above, then there are baseframes such that (Γ, ω_n) converges to the trivial group, which is linear. If the injectivity radius is unbounded below, M either has a cusp or M has arbitrarily short geodesics. A sequence of baseframes tending to a cusp yields a purely parabolic group Γ_∞ in the geometric limit. If γ is a short geodesic, then $\pi_1(M)$ appears to be almost parabolic when seen from a baseframe deep in the thin part but far from γ. Thus there is purely parabolic geometric limit in this case as well. Since any purely parabolic group is linear, (Γ, Λ) is not uniformly nonlinear.

Conversely, if the injectivity radius is bounded above and below, then any baseframe ω_n in the convex core is a bounded distance from the thick part of M. Since M is nonelementary, there are two elements $\alpha_n, \beta_n \in (\Gamma, \omega_n)$ moving ω_0 a distance bounded independent of n and generating a nonelementary group. Given any geometric limit $\Gamma_\infty = \lim (\Gamma, \omega_n)$, we can pass to a subsequence such that $\alpha_n \to \alpha$ and $\beta_n \to \beta$. A limit of discrete, nonelementary representations of the free group on two generators is still nonelementary, so Γ_∞ is nonlinear. Therefore (Γ, Λ) is uniformly nonlinear.

The lower bound $\nu_0(R_0, R_1)$ for $\nu(\Gamma, \Lambda)$ exists by compactness of the space of manifolds with given injectivity bounds and continuity of ν.

∎

Proposition 9.10 *Let $\mathcal{F}_n \to \mathcal{F}$ be a geometrically convergent sequence of saturated dynamical systems, and suppose $\Lambda_n \to \Lambda$ in the Hausdorff topology. Then $\nu(\mathcal{F}, \Lambda) \geq \liminf \nu(\mathcal{F}_n, \Lambda_n)$.*

Proof. For any ω in the convex hull of Λ, we can write $\omega = \lim \omega_n$ where ω_n is in the convex hull of Λ_n for all $n \gg 0$; then

$$\nu(\mathcal{F}, \omega) = \lim \nu(\mathcal{F}_n, \omega_n) \geq \liminf \nu(\mathcal{F}_n, \Lambda_n)$$

and the Proposition follows.

■

Rigidity. We say $\mathcal{F}$ is *rigid* if it admits no measurable invariant line field on the sphere. Equivalently, any quasiconformal conjugacy from $\mathcal{F}$ to another holomorphic dynamical system $\mathcal{F}'$ is actually conformal. Similarly, $\mathcal{F}$ is *rigid on* $\Lambda \subset \widehat{\mathbb{C}}$ if Λ supports no $\mathcal{F}$-invariant line field. The next result overlaps with Theorem 2.9.

Proposition 9.11 *If* $(\mathcal{F}, \Lambda)$ *is uniformly nonlinear, then* $\mathcal{F}$ *is rigid on* Λ.

Proof. Given a Beltrami differential μ, let (μ, ω) denote $g^*(\mu)$, where g is the unique Möbius transformation sending ω_0 to ω.

If $\mathcal{F}$ is not rigid on Λ, then there is an $\mathcal{F}$-invariant μ supported on Λ. Let $\gamma \subset \text{hull}(\Lambda)$ be a geodesic ray terminating at a point $x \in \Lambda$ where $|\mu(x)| = 1$ and μ is almost continuous. Let ω_i be a sequence of frames tending to x along γ, and equivalent up to parallel translation along γ. Then (μ, ω_i) converges weak* to a parabolic line field μ_∞. Passing to a subsequence, $(\mathcal{F}^{\text{sat}}, \omega_i)$ converges to a dynamical system $\mathcal{G}$ which preserves μ_∞. Thus $\mathcal{G}$ is linear, so $(\mathcal{F}, \Lambda)$ is not uniformly nonlinear.

■

Deformations and inflexibility. We can now adapt the arguments of §2.4 and §2.5 to prove inflexibility theorems for uniformly nonlinear and uniformly twisting dynamical systems.

A quasiconformal vector field v on $\widehat{\mathbb{C}}$ is a *deformation* of $\mathcal{F}$ if $\mu = \overline{\partial} v$ is $\mathcal{F}$-invariant. Let $S(p, r)$ denote the hyperbolic sphere of radius r about $p \in \mathbb{H}^3$, and let $Mv(p)$ denote the visual distortion of v.

Lemma 9.12 *Let $(\mathcal{F}, \Lambda)$ be uniformly nonlinear, with $\nu(\mathcal{F}, \Lambda) \geq \nu_0 > 0$. Let v be a quasiconformal deformation of $\mathcal{F}$. Then there is a radius $r = r(\nu_0) > 0$ such that whenever $S(p, r) \subset \mathrm{hull}(\Lambda)$, we have*

$$Mv(p) \leq \frac{1}{2} \sup_{q \in S(p,r)} Mv(q).$$

Proof. The proof is by contradiction.

After a conformal change of coordinates, we can assume p is the origin in the hyperbolic ball. If the Lemma is false, we can find a sequence of saturated dynamical systems $(\mathcal{F}_n, \Lambda_n)$, radii $r_n \to \infty$ and deformations v_n, such that $\nu(\mathcal{F}_n, \Lambda_n) \geq \nu_0 > 0$, $Mv_n(p) \geq 1/2$, but $Mv_n(q) \leq 1$ for all $q \in S(p, r_n) \subset \mathrm{hull}(\Lambda_n)$. After passing to a subsequence, we may assume $\mathcal{F}_n^{\mathrm{sat}} \to \mathcal{G}$ geometrically and $\Lambda_n \to \widehat{\mathbb{C}}$ in the Hausdorff topology. By Propositions 9.10 and 9.11, $(\mathcal{G}, \widehat{\mathbb{C}})$ is uniformly nonlinear and therefore $\mathcal{G}$ is rigid.

By Corollary B.18, after passing to a subsequence and correcting by conformal vector fields, we can assume v_n converges uniformly to a quasiconformal vector field v on the sphere. Since $Mv_n(p) \geq 1/2$, we have $Mv(p) \geq 1/2$. Also $\bar{\partial} v_n \to \bar{\partial} v$ as a distribution, so v too is a deformation of $\mathcal{G}$. But the rigidity of $\mathcal{G}$ implies v is conformal, so $Mv(p) = 0$, a contradiction.

∎

Iterating this bound, as in Theorem 2.15, we obtain:

Corollary 9.13 (Infinitesimal inflexibility) *Let $V = \mathrm{ex}(v)$ be the visual extension of a quasiconformal deformation v of $(\mathcal{F}, \Lambda)$. Then for any $p \in \mathbb{H}^3$ we have:*

$$\|SV(p)\| \leq CMv(p) \leq C' \exp(-\alpha\, d(p, \mathbb{H}^3 - \mathrm{hull}(\Lambda))) \|Sv\|_\infty,$$

where $\alpha > 0$ depends only on $\nu(\mathcal{F}, \Lambda) > 0$.

The constants C and C' are universal.

9.3   Uniform twisting

In this section we complete the proof of a *dynamic* inflexibility theorem for conjugacies, analogous to the geometric inflexibility Theorems 2.11 and 2.18.

Definitions. A dynamical system $\mathcal{F} \subset \mathcal{V}(\widehat{\mathbb{C}} \times \widehat{\mathbb{C}})$ is *twisting* if any $\mathcal{F}'$ quasiconformally conjugate to $\mathcal{F}$ is nonlinear. A collection of dynamical systems $\{\mathcal{F}_\alpha\}$ is *uniformly twisting* if $\lim \mathcal{F}^{\mathrm{sat}}_{\alpha(i)}$ is twisting for any geometrically convergent sequence in the family of saturations.

Examples. The group $\Gamma_\lambda = \langle z \mapsto \lambda^n z \; : \; n \in \mathbb{Z} \rangle$ is linear if $\lambda \in \mathbb{R}$ (since it preserves $\mu = d\bar{z}/dz$), and twisting if $\lambda \in S^1 - \{\pm 1\}$. If $|\lambda| \neq 1$, then Γ_λ is quasiconformally conjugate to a group with $\lambda \in \mathbb{R}$, so it is *not* twisting. Any nonelementary Kleinian group is twisting, as is any rational map of degree $d > 1$. In fact, a holomorphic mapping $f : U \to V$ with a critical point in U is never conjugate to a linear mapping, so any holomorphic dynamical system $\mathcal{F}$ with a critical point is twisting.

In applications it is often just as easy to show $(\mathcal{F}, \Lambda)$ is uniformly twisting as to show it is uniformly nonlinear. For example, the proof of Proposition 9.9 also shows (Γ, Λ) is uniformly twisting, since all the geometric limits obtained were nonelementary.

Unlike nonlinearity, twisting is preserved under quasiconformal deformation. This property is a substitute, in the dynamical setting, for the fact that the injectivity radius of a hyperbolic manifold changes by only a bounded factor under quasi-isometry.

For $K \geq 1$, let

$$\nu^K(\mathcal{F}) = \inf_\phi \nu(\phi_*(\mathcal{F})),$$

where the infimum is over K-quasiconformal mappings $\phi : \widehat{\mathbb{C}} \to \widehat{\mathbb{C}}$, fixing 0, 1 and ∞ and conjugating $\mathcal{F}$ to another holomorphic dynamical system $\phi_*(\mathcal{F})$. Set

$$\nu^K(\mathcal{F}, \Lambda) = \inf_{\text{frames } \omega \text{ in hull}(\Lambda)} \nu^K(\mathcal{F}, \omega).$$

Then $(\mathcal{F}, \Lambda)$ is uniformly twisting if and only if $\nu^K(\mathcal{F}, \Lambda) > 0$ for all K.

Proposition 9.14 *Suppose* $(\mathcal{F}, \Lambda)$ *is uniformly twisting, and let* $\phi : \widehat{\mathbb{C}} \to \widehat{\mathbb{C}}$ *be a* K-*quasiconformal map. Then* $(\phi_*(\mathcal{F}), \phi(\Lambda))$ *is also uniformly twisting, and*

$$\nu(\phi_*(\mathcal{F}), \phi(\Lambda)) \geq \nu_0 > 0$$

where ν_0 *depends only on* K *and* $\nu^K(\mathcal{F}, \Lambda)$.

Proof. First we remark that a bounded change in basepoint makes a controlled change in nonlinearity. That is, if $(\mathcal{F}, \omega_1)$ is nonlinear and ω_2 is a frame based at distance less than r from ω_1, then $\nu(\mathcal{F}, \omega_2) \geq \epsilon > 0$, where ϵ depends only on $\nu(\mathcal{F}, \omega_1)$ and r.

Returning to the Proposition, let $\Phi : \mathbb{H}^3 \to \mathbb{H}^3$ be an L-quasi-isometry extending ϕ; we may assume $L \leq K^{3/2}$ (see Corollary B.23). Consider a frame ω_1 over a point p_1 in hull$(\phi(\Lambda))$. By Proposition 2.16, there is a $p_0 \in$ hull(Λ) such that $d(p_1, \Phi(p_0))$ is bounded in terms of K. Up to a change of coordinates for $\mathcal{F}$, we can assume p_0 is the origin in the hyperbolic ball. After changing ϕ by post-composition with a Möbius transformation, we can also assume that ϕ fixes 0, 1 and ∞. Then by definition,

$$\nu(\phi_*(\mathcal{F}), \omega_0) \geq \nu^K(\mathcal{F}, \Lambda).$$

Because of our normalization of ϕ to fix three points, $d(p_0, \Phi(p_0))$ is also bounded in terms of K. Thus $d(p_0, p_1) \leq d_0(K)$, so by our initial remarks $\nu(\phi_*(\mathcal{F}), \omega_1) \geq \nu_0 > 0$, where ν_0 depends only on $d_0(K)$ and $\nu^K(\mathcal{F}, \Lambda)$.

Since ω_1 was arbitrary, we have $\nu(\phi_*(\mathcal{F}), \phi(\Lambda)) \geq \nu_0 > 0$. This shows $(\phi_*(\mathcal{F}), \phi(\Lambda))$ is uniformly nonlinear. By similar considerations, it is also uniformly twisting.

∎

Theorem 9.15 (Dynamic inflexibility) *Let* $(\mathcal{F}, \Lambda)$ *be uniformly twisting, and let* $\phi : \widehat{\mathbb{C}} \to \widehat{\mathbb{C}}$ *be a* K-*quasiconformal conjugacy from* $\mathcal{F}$ *to another holomorphic dynamical system* $\mathcal{F}'$. *Then:*

1. *there is a natural extension of* ϕ *to a volume-preserving diffeomorphism* $\Phi : \mathbb{H}^3 \to \mathbb{H}^3$ *whose pointwise quasi-isometry constant satisfies*

$$L(\Phi, p) \leq 1 + C \exp(-\alpha\, d(p, \partial H))$$

for all $p \in H = \text{hull}(\Lambda)$; and

2. *for any δ-deep point x of Λ, ϕ is $C^{1+\beta}$-conformal at x.*

The constants C and $\alpha, \beta > 0$ depend only on K, δ and $\nu^K(\mathcal{F}, \Lambda)$.

Because injectivity bounds in the setting of Kleinian groups are the same as the assumption of uniform twisting, this result generalizes Theorems 2.11 and 2.18.

Sketch of the proof. The proofs of these two statements are almost identical to the ones for Kleinian groups, so we will not repeat all the details. As before, we may include ϕ in a Beltrami isotopy ϕ_t which extends to a quasi-isometric isotopy Φ_t of hyperbolic space. These isotopies are generated by vector fields v_t and V_t respectively.

The main new point is that uniform twisting gives the analogue of injectivity bounds during the course of the isotopy. That is, for $(\mathcal{F}_t, \Lambda_t) = ((\phi_t)_*(\mathcal{F}), \phi_t(\Lambda))$, Proposition 9.14 guarantees $\nu(\mathcal{F}_t, \Lambda_t) \geq \nu_0 > 0$ for all t. Then Corollary 9.13 shows the strain SV_t and the visual distortion Mv_t decay exponentially fast in the convex hull of Λ_t, with an estimate that is independent of t. By Proposition 2.16, $d(\Phi_t(p), \partial H)$ does not change too much during isotopy, so the integrated map Φ_1 is close to an isometry when p is deep in the convex hull. This completes the proof of the bounds on $L(\Phi, p)$.

Conformality of ϕ at deep points follows from bounds on the visual distortion and Theorem B.26, exactly as in the proof of Theorem 2.18.

■

Relative inflexibility. One often encounters deformations which leave the conformal structure fixed on part of the sphere. In this case, one may adjoin the region Ω_0 of conformality to the set Λ_0 where the dynamics is twisting, to obtain a rigidity theorem on their union Λ.

Theorem 9.16 (Relative inflexibility) *Let $\Lambda = \Lambda_0 \cup \Omega_0 \subset \hat{\mathbb{C}}$, where Λ and Λ_0 are closed and Ω_0 is open. Let $(\mathcal{F}, \Lambda_0)$ be uniformly twisting, and let $\phi : \hat{\mathbb{C}} \to \hat{\mathbb{C}}$ be a K-quasiconformal conjugacy from $\mathcal{F}$ to $\mathcal{F}'$ which is conformal on Ω_0.*

Then the conclusions of Theorem 9.15 hold, with constants depending only on K, δ and $\nu^K(\mathcal{F}, \Lambda_0)$.

The proof is a straightforward generalization of that of Theorem 9.15.

9.4 Quadratic maps and universality

In this section we describe how towers fit in with the notion of geometric limits. Then we show certain infinitely renormalizable maps are uniformly twisting on their Julia sets. By the inflexibility theory just presented, quasiconformal conjugacies between such mappings are $C^{1+\alpha}$-conformal at the critical point.

In particular, when the inner class of f is fixed by $\mathcal{R}_p$ (as is the case for the Feigenbaum polynomial), any quasiconformal conjugacy ϕ between f^p and f is is differentiable at the origin, and its derivative $\phi'(0) = \alpha$ is universal.

Exponential convergence of renormalization follows easily from differentiability of conjugacies, as we will see in the next section.

The full dynamics $\mathcal{F}(f)$. Let $f : U \to V$ be a holomorphic map between open sets in the Riemann sphere. The iterates $f^i(z)$ generally have smaller domains than f itself. Let $\mathrm{gr}(f^{-i} \circ f^j)$ denote the relation consisting of all pairs (z, w) such that $f^j(z) = f^i(w)$.

Let $\mathcal{F}(f)$ denote the holomorphic dynamical system consisting of the graphs of all holomorphic maps $g : U' \to V'$ such that $\mathrm{gr}(g) \subset \mathrm{gr}(f^{-i} \circ f^j)$ for some $i, j \geq 0$. We similarly define $\mathcal{F}^+(f)$ by the requirement $\mathrm{gr}(g) \subset \mathrm{gr}(f^i)$ for some $i \geq 0$. It is straightforward to see that f, $\mathcal{F}(f)$, and $\mathcal{F}^+(f)$ have the same quasiconformal deformations (any Beltrami differential invariant by one is invariant by the others).

Remark. The space $\mathcal{F}(f)$ may contain maps which are not in the saturation of $\{\mathrm{gr}(f^{-i} \circ f^j) : i, j \geq 0\}$. For example, when $f(z) = z^2$, the map $g(z) = -z$ belongs to $\mathcal{F}(f)$, but for $i > 0$ the hypersurface $\mathrm{gr}(f^{-i} \circ f^i)$ contains irreducible components other than $\mathrm{gr}(g)$ passing through $(0, 0)$.

When f is infinitely renormalizable, $\mathcal{F}^+(f)$ contains all the quadratic-like maps $f^n : U_n \to V_n$, $n \in \mathcal{SR}(f)$. So by compactness of towers, it is straightforward to check:

Proposition 9.17 *Let $f(z) = z^2 + c$ be infinitely renormalizable, with bounded combinatorics and definite moduli bounded by (m, B).*

Let $\mathcal{F} = \overline{\mathcal{F}^+(f)}$, and let $A_n(z) = \alpha_n z$ where $\alpha_n \to \infty$. Then after passing to a subsequence, $(A_n)_*(\mathcal{F})$ converges geometrically, and there is a bi-infinite tower

$$\mathcal{T} = \langle f_s; s \in S \rangle \in \mathcal{T}ow_0^\infty(m, B)$$

such that

$$\lim (A_n)_*(\mathcal{F}) \supset \{\mathrm{gr}(f_s) : s \in S\}.$$

Thus the theory of towers captures a piece of the geometric limit of f blown up at its critical point. In particular, any such geometric limit is rigid, by rigidity of towers (Theorem 6.1). We only need the forward dynamics of f to obtain this rigidity.

Next we examine geometric limits about general points in the Julia set of f. These limits are not always rigid, but if we consider the full dynamics of f, they are at least twisting.

Theorem 9.18 *Let $f(z) = z^2 + c$ be infinitely renormalizable, with bounded combinatorics and definite moduli. Then $(\mathcal{F}(f), J(f))$ is uniformly twisting.*

Proof. Uniform twisting follows from Theorem 8.10 (Small Julia sets everywhere). Suppose ω_n is a sequence of baseframes in the convex hull of the Julia set, based at points (z_n, t_n) in the upper half-space $\mathbb{H}^3 = \mathbb{C} \times \mathbb{R}_+$. Then the ball $B(z_n, t_n) \subset \mathbb{C}$ meets the Julia set (otherwise (z_n, t_n) would be separated from $J(f)$ by a half-plane).

By Theorem 8.10, there is a quadratic-like map $g_n : U_n \to V_n$ in $\mathcal{F}(f)$ with $\mathrm{mod}(V_n, U_n) > m' > 0$, $d(z_n, J(g_n)) = O(t_n)$ and $\mathrm{diam}\, J(g_n) \asymp t_n$; here the constants depend only on (m, B). Applying an affine Möbius transformation to move ω_n to a frame at the basepoint $(0, 1) \in \mathbb{H}^3$, we obtain a new polynomial-like map h_n with $\mathrm{diam}\, J(h_n) \asymp 1$ and $d(0, J(h_n)) = O(1)$. Any two frames at the basepoint differ by a rotation, so h_n belongs to $(\mathcal{F}(f), \omega_n)$ up to a rotation. Thus any geometric limit of the form $\mathcal{G} = \lim(\mathcal{F}(f)^{\mathrm{sat}}, \omega_n)$ contains the graph of a limiting quadratic-like map (possibly conjugated by a rotation of the sphere), so it is twisting. Therefore $(\mathcal{F}(f), J(f))$ is uniformly twisting.

∎

The proof yields a somewhat more precise statement:

Theorem 9.19 *Let $f(z) = z^2 + c$ be infinitely renormalizable, with combinatorics and moduli bounded by (m, B), and suppose $f : U \to V$ is a quadratic-like restriction of f with $\mathrm{mod}(U, V) \geq m > 0$. Then*

$$\nu^K(\mathcal{F}(f|U), J(f)) \geq \nu_0(K, m, B) > 0.$$

Proof. The geometric limits of $\mathcal{F}(f|U)$ contain the same quadratic-like maps as in the preceding proof, except with possibly smaller moduli (still bounded below in terms of (m, B)). Because of the critical point, these quadratic-like limits retain definite nonlinearity under a quasiconformal deformation.

■

Corollary 9.20 *Let f and g be infinitely renormalizable quadratic-like maps with the same inner class, and with combinatorics and moduli bounded by (m, B). Then there is a quasiconformal map $\phi : \mathbb{C} \to \mathbb{C}$ which is a conjugacy between f and g near their Julia sets, and which is $C^{1+\alpha}$-conformal at the critical point, where α depends only on (m, B).*

Proof. It suffices to treat the case where $f(z) = z^2 + c$, $c = I(g)$. By Theorem 8.8, the critical point $z = 0$ is a $\delta(m, B)$-deep point of $J(f)$. By Proposition 4.6, there is a $K(m)$-quasiconformal map $\phi : \mathbb{C} \to \mathbb{C}$ conjugating $f : U \to V$ to $g : U' \to V'$, where $\mathrm{mod}(U, V)$ and $\mathrm{mod}(U', V')$ are at both at least $m/2$. By the preceding result, $\nu^{K(m)}(\mathcal{F}(f|U), J(f))$ is bounded below in terms of (m, B), so by Theorem 9.15 ϕ is $C^{1+\alpha(m,B)}$-conformal at $z = 0$.

■

Universality. Finally consider a real quadratic-like map $f : U \to V$ with a periodic tuning invariant (such as the Feigenbaum polynomial), normalized so $f'(0) = 0$. Equivalently, assume the inner class $I(f) \in \mathbb{R}$ is fixed by $\mathcal{R}_p$ for some p. Then there is a quasiconformal conjugacy $\phi : V_p \to V$ between $f : U \to V$ and a renormalization $f^p : U_p \to V_p$. Note $\phi(0) = 0$ since ϕ preserves the critical point.

Following the example of totally degenerate groups discussed in §3.7, we can now make several parallel assertions about the infinitely renormalizable mapping f.

1. *The critical point $z = 0$ is a deep point of $J(f)$.* Indeed, this property is proved in §8 for quadratic polynomials, and it is inherited by f via hybrid conjugacy.

2. *The map ϕ is an endomorphism of the dynamics of f.* That is, $\phi \circ f^p = f \circ \phi$.

3. *The map ϕ is $C^{1+\alpha}$-conformal at all deep points of the small Julia set $J_p(f)$.* This follows from Theorems 7.15, 9.15 and 9.19.

4. *We have $|\phi'(0)| > 1$.*

5. *The self-similarity factor $\phi'(0)$ is universal, in the sense that it only depends on the combinatorics of f.* To see these last two statements, recall $\mathcal{R}_p^n(f) \to F$ in $\mathcal{G}$, where F is the unique fixed point of renormalization with the same combinatorics as f. Under rescaling, $\phi(z)$ converges to the linear map $z \mapsto \alpha z$ where $F^p(z) = \alpha^{-1}F(\alpha z)$. Therefore $\phi'(0) = \alpha$ depends only on the combinatorics of f, and $|\alpha| > 1$.

In summary, the fine structure of f at the critical point is uniquely determined by its combinatorics.

9.5 Speed of convergence of renormalization

In this section we show that renormalization converges exponentially fast in a fixed inner class.

Theorem 9.21 (Rapid convergence of renormalization) *Let $\tilde{F} : W \to \mathbb{C}$ be a fixed point of $\mathcal{R}_p$, $p > 1$. Then there is a constant $\lambda < 1$ such that for any $f \in \mathcal{H}$ with the same inner class as $\tilde{F}$, and any compact $K \subset W$, we have*

$$\sup_{z \in K} |\mathcal{R}_p^n(f)(z) - \tilde{F}(z)| \leq \lambda^n$$

for all n sufficiently large.

Proof. Let α be the renormalization factor for $\widetilde{F}$, so $|\alpha| > 1$ and $\widetilde{F}(z) = \alpha \widetilde{F}^p(\alpha^{-1} z)$. A quadratic-like restriction $F : U \to V$ of $\widetilde{F}$ has combinatorics bounded by $B = p$. We claim F has definite moduli. Indeed, since $\mathcal{SR}(f)$ is ordered by divisibility and contains all powers of p, there are finitely many a_i such that any $n \in \mathcal{SR}(f)$ can be expressed in the form $n = p^k a_i$. Since $\mathcal{R}_p^n(F) \to \widetilde{F}$, we have

$$\liminf_k \, \mathrm{mod}(\mathcal{R}_{p^k a_i}(F)) \;\geq\; \mathrm{mod}(\mathcal{R}_{a_i}(\widetilde{F})) > 0$$

for each i; and therefore $m = \inf_{\mathcal{SR}(F)} \mathrm{mod}(\mathcal{R}_n(F)) > 0$.

By Theorem 7.13, we also have $\mathcal{R}_p^n(f) \to \widetilde{F}$ uniformly on compact subsets of W. So replacing f with a quadratic-like restriction of $\mathcal{R}_p^n(f)$ for $n \gg 0$, we can assume $\inf_{\mathcal{SR}(f)} \mathrm{mod}(\mathcal{R}_n(f)) > m/2$.

By Corollary 9.20, there is a conjugacy ϕ from F to f defined near $J(F)$ which is $C^{1+\delta}$-conformal at the critical point $z = 0$. The exponent δ depends only on (m, B), and thus only on $\widetilde{F}$. We will prove that the Theorem holds for any λ with $1 > \lambda > \alpha^{-\delta} > 0$.

Conjugating f by a dilation (which does not affect the sequence of renormalizations), we can assume the conjugacy ϕ from F to f satisfies $\phi'(0) = 1$. Then $\phi(z) = z + O(|z|^{1+\delta})$ for all z near 0.

Let $\beta_n \to 0$ denote the β fixed points of f^{p^n} in $J_{p^n}(f)$. The β fixed points in $J_{p^n}(F)$ are at α^{-n}. Since $\phi(\alpha^{-n}) = \beta_n$, the ratios of corresponding fixed points satisfy

$$\rho_n \;=\; \frac{\alpha^{-n}}{\beta_n} \;=\; 1 + O(\alpha^{-n\delta}) = 1 + o(\lambda^n).$$

Similarly $\phi_n = \alpha^n \phi \alpha^{-n}$ satisfies

$$\phi_n(z) = z + o(\lambda^n)$$

uniformly on compact subsets of $\mathbb{C}$. The map $\rho_n \phi_n$ conjugates $\mathcal{R}_p^n(F)$ to $\mathcal{R}_p^n(f)$, so for all n large enough we have

$$|\mathcal{R}_p^n(F)(z) - \mathcal{R}_p^n(f)(z)| = o(\lambda^n)$$

for all z in K. But $\mathcal{R}_p^n(F) = \widetilde{F}$ on K, completing the proof.

∎

The main ingredients in this proof — the differentiability of conjugacies and the exponential shrinking of Julia sets — are present for any quadratic-like map with bounded combinatorics and definite moduli. Thus we have:

Theorem 9.22 *Let f and g be a pair of infinitely renormalizable quadratic-like maps in $\mathcal{H}$ with the same inner class. Suppose f has bounded combinatorics and definite moduli. Then*

$$|\mathcal{R}_n(f)(z) - \mathcal{R}_n(g)(z)| \leq n^{-\gamma}$$

for all sufficiently large $n \in \mathcal{SR}(f)$ and for all z in an ϵ-neighborhood of $J(\mathcal{R}_n(f))$. The constants $\gamma, \epsilon > 0$ depend only on bounds (m, B) for f.

Note: If $\mathcal{SR}(f) = \{n_0, n_1, n_2, \dots\}$, then $n_i \geq 2^i$, so $n_i^{-\gamma} \leq i^{-\lambda}$, where $\lambda = 2^{-\gamma}$. Thus the bound above is exponential in the *number of renormalizations i*.

Proof. Since f and g are quasiconformally conjugate, g also has bounded combinatorics and definite moduli. So by Corollary 9.20, there is a conjugacy between f and g near their Julia sets which is differentiable at $z = 0$. Thus for large n, $\mathcal{R}_p^n(f)$ and $\mathcal{R}_p^n(g)$ are uniformly close near their filled Julia sets. By the pullback argument (§4.3), we can obtain a nearly conformal conjugacy ϕ between them. Replacing f and g with quadratic-like restrictions of their high renormalizations, we can assume both have moduli bounded below by $m/2$. Applying Corollary 9.20 again, we obtain a new conjugacy ϕ between f and g such that ϕ is $C^{1+\alpha(m,B)}$-conformal at $z = 0$.

By Proposition 8.1, $\operatorname{diam} J_n(f) = O(n^{-\beta(m,B)})$. Therefore ϕ gives rise to a conjugacy between $\mathcal{R}_n(f)$ and $\mathcal{R}_n(g)$ which is within $O(n^{-\alpha\beta})$ of the identity near their Julia sets. Since both f and g have definite moduli, an $\epsilon(m)$-neighborhood of $J(\mathcal{R}_n(f))$ lies in the domain of $\mathcal{R}_n(f)$. Thus the Theorem holds for any γ with $\alpha\beta > \gamma > 0$.

∎

Corollary 9.23 *Let f and g be infinitely renormalizable real symmetric quadratic-like maps with combinatorics bounded by B. Suppose f and g have the same tuning invariant. Then*

$$\sup_{x\in[-1,1]} |\mathcal{R}_n(f)(x) - \mathcal{R}_n(g)(x)| \leq n^{-\gamma(B)}$$

for all $n \in \mathcal{SR}(f)$ sufficiently large, where $\gamma(B) > 0$.

Proof. By Sullivan's Theorem 7.15, f and g have the same inner class c, and $h(z) = z^2 + c$ has moduli bounded below by $m(B) > 0$. By the preceding Theorem, $\mathcal{R}_n(f)$ and $\mathcal{R}_n(g)$ both converge rapidly to $\mathcal{R}_n(h)$ on $[-1, 1]$, since $[-1, 1] \subset J(\mathcal{R}_n(h))$. The Corollary follows by the triangle inequality.

∎

Corollary 9.24 *There is a $C^{1+\alpha(B)}$ map $\phi : \mathbb{R} \to \mathbb{R}$ providing a conjugacy between f and g on their postcritical sets.*

This Corollary follows from the preceding one; a proof can be given using the scale function of the Cantor set $P(f)$. See [MeSt, Theorem 9.4].

10 Conclusion

In this chapter we briefly recapitulate the analogies between the action of a mapping class $\psi : AH(S) \to AH(S)$ and the action of a renormalization operator $\mathcal{R}_p : \mathcal{H}^{(p)} \to \mathcal{H}$. We also discuss some open problems on both sides of the dictionary. These parallels and problems were summarized in Tables 1.1 and 1.2 of §1.

Consider first the algorithms for the construction of fixed points, detailed in §3.4 for ψ and in §7.1 for $\mathcal{R}_p$. In each case, the dynamics at the fixed point seems to be hyperbolic, with both expanding and contracting directions. The fixed point cannot be located by iterating a generic starting point; for example, $\psi^n(M) \to \infty$ in $AH(S)$ if M is a quasifuchsian group, and $\mathcal{R}_p^n(f)$ eventually leaves the domain $\mathcal{H}^{(p)}$ of $\mathcal{R}_p$ for a generic quadratic-like map f.

Instead, to locate the fixed point, we first construct a point on its stable manifold. See Figures 10.1 and 10.2.

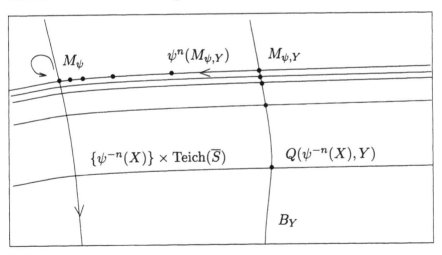

Figure 10.1. Dynamics of ψ on $AH(S)$.

In the Kleinian group setting, the quasifuchsian subspace $QF(S)$ of $AH(S)$ has a product structure

$$QF(S) = \text{Teich}(S) \times \text{Teich}(\overline{S})$$

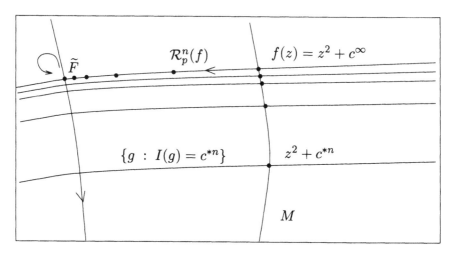

Figure 10.2. Dynamics of $\mathcal{R}_p$ on $\mathcal{H}$.

which is preserved by ψ. In particular, ψ preserves the foliation whose leaves have the form $\{X\} \times \mathrm{Teich}(\overline{S})$. Therefore ψ acts on the space of leaves. A Bers slice B_Y cuts each leaf in a single point, so this dynamics on the space of leaves can be represented by the map of B_Y to itself given by $Q(X, Y) \mapsto Q(\psi(X), Y)$. The point

$$M_{\psi, Y} = \lim Q(\psi^{-n}(X), Y)$$

on the stable manifold of ψ is located by iterating the *inverse* of this induced mapping on the transversal B_Y. Once $M_{\psi, Y}$ is constructed, the fixed point M_ψ is simply the limit of $\psi^n(M_{\psi, Y})$.

In the case of extended quadratic-like maps, the foliation of $\mathcal{H}$ by the level sets of the inner class is preserved by $\mathcal{R}_p$. If we embed the Mandelbrot set M into $\mathcal{H}$ by sending c to $z^2 + c$, then M provides a transversal to the leaves of *this* foliation. The induced dynamics on M is many-to-one; the tuning map $x \mapsto c * x$ is a particular inverse branch of the dynamics. Iterated tuning converges (at least for $c \in \mathbb{R}$) to the point $f(z) = z^2 + c^\infty$ on the stable manifold of $\mathcal{R}_p$, and then the fixed point $\widetilde{F}$ is simply the limit of $\mathcal{R}_p^n(f)$.

In both cases, the objects in a given leaf of the foliation lie in a single quasiconformal conjugacy class.

The first stage of the construction yields a special geometrically infinite object (a totally degenerate group $M_{\psi,Y}$ or an infinitely renormalizable map $z^2 + c^\infty$) as a limit of geometrically finite ones (quasifuchsian groups or critically finite maps). In the second stage (iteration of ψ or $\mathcal{R}_p$), the dynamical system does not change: we simply view it from a varying perspective, by changing the choice of generators for the group, or by renormalizing the mapping.

The objects $M_{\psi,Y}$ and f each admit an approximate symmetry of the desired type: ψ can be realized by a quasi-isometry on $M_{\psi,Y}$, and f is quasiconformally conjugate to f^p. By re-marking or renormalization, one passes deeper into the structure of these geometrically infinite objects. The deep structure is more rigid, and in the limit the approximate symmetry becomes exact (isometric or conformal).

To show the symmetry becomes exact, it suffices to establish rigidity of geometric limits. In the hyperbolic setting we have used rigidity of manifolds with injectivity radius bounded above; for renormalization, the rigidity of towers. Note that the quadratic-like germ $[F]$ of the fixed point of renormalization is *not* quasiconformally rigid in its own right; its rigidity comes about when it is embedded in the tower $\mathcal{T}_\infty$, which records that fact that F comes from high renormalizations of f.

Open problems. In conclusion we record some directions for further study.

1. Do the iterated tunings c^{*n} converge when c is a strictly complex superstable point in the Mandelbrot set?

2. Do the renormalizations of the limiting map $z^2 + c^\infty$ have definite moduli?

3. Let $f(z) = z^2 + c$ be an infinitely renormalizable mapping with bounded combinatorics and definite moduli. Does the Julia set of f have Hausdorff dimension two? Is its area zero?

   It is known that the limit set of $M_{\psi,Y}$ has Hausdorff dimension two and measure zero, where $M_{\psi,Y}$ is the totally degenerate point on the boundary of a Bers slice arising as a limit of $Q(\psi^{-n}(X), Y)$. The dimension two result was proved by Sullivan in the case where S is closed [Sul2]; more generally, Bishop and Jones have shown the limit set of any geometrically infinite, finitely generated Kleinian group has Hausdorff dimension two [BJ]. The measure zero result was established by Thurston [Th1, §8]; see also [Bon]. On the other hand, examples of polynomials with Julia sets of positive area have been announced by Nowicki and van Strien [NS].

4. Let $\mathcal{T}$ be a bi-infinite tower with bounded combinatorics and definite moduli. Does $\mathcal{T}$ act ergodically on $\mathbb{C}$? That is, if $A \subset \mathbb{C}$ is a set of positive measure such that $f_s^{-1}(A) \subset A$ for all maps f_s in $\mathcal{T}$, does $A = \mathbb{C}$ a.e. ?

   This question is suggested by the known ergodicity of the action of the fundamental group of M_ψ on $\widehat{\mathbb{C}}$, also established in [Th1, §8].

5. What is the analogue of bounded combinatorics in the setting of Kleinian groups?

   For example, much of the theory discussed for ψ^n should work for an arbitrary sequence $\psi_n \in \mathrm{Mod}(S)$, subject to the condition that the geodesic segments from X to $\psi_n(X)$ in $\mathrm{Teich}(S)$ lie over a compact set in moduli space. When S is a punctured torus, this condition is related to continued fractions of bounded type. What property of ψ_n corresponds to this Diophantine condition on higher genus surfaces?

6. Is the Mandelbrot set self-similar about c^∞?

   The self-similarity of the real bifurcation locus about the Feigenbaum point is one of the basic renormalization conjectures. The self-similarity factor is conjecturally equal to the expanding eigenvalue $\lambda = 4.66920\ldots$ of the renormalization operator $\mathcal{R}_2$ at its fixed point.

   See Figure 10.3 for two blowups of the Mandelbrot set about the Feigenbaum point. The conjectural self-similarity of M about generalized Feigenbaum points is discussed in detail in [Mil].

7. Is the boundary of a Bers slice B_Y self-similar about $M_{\psi,Y}$?

   Self-similarity of the boundary of Teichmüller space about limits of pseudo-Anosov mappings was observed in computer investigations by Wright and the author in 1989. The particular

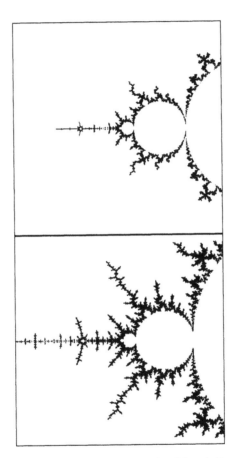

Figure 10.3. Self-similarity in the Mandelbrot set.

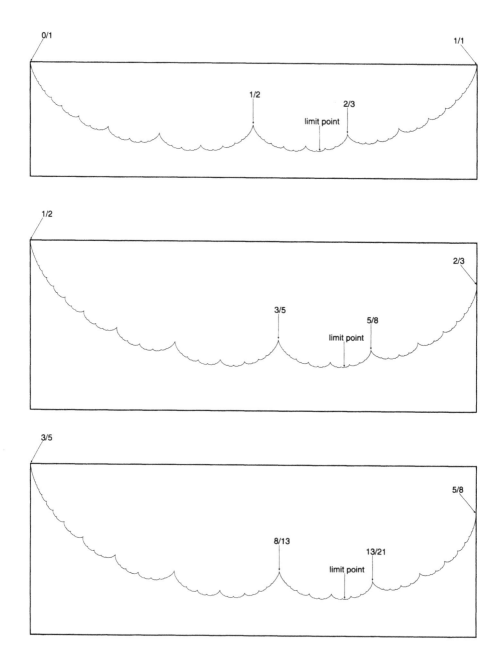

Figure 10.4. Self-similarity in the boundary of Teichmüller space.

case we studied was a Maskit slice of $AH(S)$, where S is a torus with one boundary component. A Maskit slice is like a Bers slice, except that instead of holding one conformal structure fixed, a particular simple curve γ is pinched to a cusp. (This accidental parabolic makes the Maskit slice easier to compute with.)

The modular group of a punctured torus is $SL_2(\mathbb{Z})$, and the simplest pseudo-Anosov mapping class is $\psi = \left(\begin{smallmatrix} 2 & 1 \\ 1 & 1 \end{smallmatrix}\right)$. The corresponding point $M_{\psi,\gamma}$ in the boundary of a Maskit slice can be computed as a limit of cusp groups which are naturally labeled by ratios of successive Fibonacci numbers.

Figure 10.4 depicts a single period of the Maskit boundary, together with two blowups around the limit point $M_{\psi,\gamma}$. The boundary appears to scale by $\lambda = 4.79129\ldots = (5 + \sqrt{21})/2$, which is also the expanding eigenvalue of ψ at its fixed point M_ψ in $AH(S)$. This example appeared in [Mc3]; for more details, see [Wr], [MMW]. The pictures were drawn by Wright's computer program, which uses Newton's method and Farey fractions to enumerate cusp points on the boundary.

8. Can the ideas of towers, deep points and uniform twisting shed light on other infinitely renormalizable dynamical systems? Some steps in this direction appear in [NS] for the Fibonacci kneading sequence, using Lyubich's generalized quadratic-like maps. For critical circle mappings, *a priori* bounds are available by work of de Faria [dF], and new insights are under development by de Faria and de Melo [dFdM]. Applications to Siegel disks in the quadratic family appear in [Mc7]. Other approaches to renormalization contraction are also under development; see for example [Ka].

Appendix A

Quasiconformal maps and flows

In this Appendix we discuss quasiconformal maps and flows in n-dimensional space. The main result we will develop is:

Theorem A.1 (Quasiconformal integrability) *Let $v : S^n \times [0,1] \to TS^n$ be a continuous, time-dependent quasiconformal vector field on the n-sphere, $n > 1$, with $\|Sv_t\|_\infty \le k$ for all t.*

Then there is a unique isotopy $\phi : S^n \times [0,1] \to S^n$ with $\phi_0(x) = x$ and

$$\frac{d\phi_t}{dt}(x) = v_t(\phi_t(x)).$$

The map ϕ_t is K_t-quasiconformal, where $K_t = \exp(ntk)$.

This theorem is due to Reimann [Rei2]; see also [Ah3]. The tensor Sv_t measures the quasiconformal distortion of the vector field v_t, as described below.

We begin with an intrinsic discussion of conformal structures and their deformations. Then we prove a Sobolev-type result: a function with derivatives in BMO satisfies the Zygmund condition. The Zygmund condition insures that a quasiconformal vector field has an $x \log x$ modulus of continuity, which implies unique integrability. The Theorem then follows by smoothing and passing to a limit. Several basic estimates and compactness results for quasiconformal vector fields are also derived.

A.1   Conformal structures on vector spaces

Fix $n > 1$, and let V be an n-dimensional vector space over $\mathbb{R}$. A *metric g* on V is a positive symmetric quadratic form. More precisely, we regard a metric as an isomorphism $g : V \to V^*$ such that $g^* = g$ and $\langle g(v), v \rangle \ge 0$. The associated norm is given by $|v|_g^2 = \langle g(v), v \rangle$.

A *conformal structure* on V is the choice of a metric up to scale.

Given a linear isomorphism $T : (V, g) \to (W, h)$ between metrized vector spaces, the *quasi-isometry constant* $L(T)$ is the least $L \ge 1$

such that

$$\frac{1}{L}|v|_g \leq |Tv|_h \leq L|v|_g$$

for all $v \in V$. The image of the unit sphere under T lies between spheres of radii L and $1/L$.

The conformal distortion of T is measured by the *dilatation* $K(T)$, defined to be the least $K \geq 1$ such that

$$\frac{1}{K}|\det T| \;\leq\; \left(\frac{|Tv|_h}{|v|_g}\right)^n \;\leq\; K|\det T|$$

for all $v \neq 0$. Here $|\det T|$ is measured using g and h. If T is volume-preserving, then $K(T) = L(T)^n$.

It is easy to see $K(ST) \leq K(S)K(T)$, $L(ST) \leq L(S)L(T)$, and the conditions $K(T) = 1$ and $L(T) = 1$ characterize conformal and isometric linear maps respectively.

The space of conformal structures. Let $\mathrm{Conf}(V)$ denote the space of all conformal structures on V. The group $\mathrm{GL}(V)$ of automorphisms $T : V \to V$ acts transitively on $\mathrm{Conf}(V)$ on the right, by $g \cdot T = T^*gT$.

For any fixed g we may identify $\mathrm{Conf}(V)$ with the homogeneous space

$$\mathrm{Conf}(V) = (\mathbb{R}^* \, \mathrm{O}(V, g)) \backslash \mathrm{GL}(V) = \mathrm{SO}(V, g) \backslash \mathrm{SL}(V)$$

where $\mathbb{R}^*$ is the center of $\mathrm{GL}(V)$, $\mathrm{SL}(V)$ is the group of automorphisms of determinant one, and $\mathrm{SO}(V, g)$ is the group of orientation-preserving isometries of g.

The *Teichmüller metric* on $\mathrm{Conf}(V)$ is defined by

$$d(g, h) \;=\; \frac{1}{n} \log K(I : (V, g) \to (V, h)),$$

where I is the identity map.

Deformations. We now pass to the level of Lie algebras to describe deformations of conformal structures. The Lie algebra $\mathrm{gl}(V)$ may be identified with the space of endomorphisms $A : V \to V$. Let

$$\begin{aligned}
\mathrm{sym}_0(V, g) &= \{A \in \mathrm{gl}(V) \,:\, A^*g = gA \text{ and } \mathrm{tr}(A) = 0\}, \text{ and let} \\
\mathrm{so}(V, g) &= \{A \in \mathrm{gl}(V) \,:\, A^*g = -gA\}.
\end{aligned}$$

Then
$$\mathrm{gl}(V) \;=\; \mathbb{R} \oplus \mathrm{so}(V,g) \oplus \mathrm{sym}_0(V,g)$$

is a direct sum decomposition of $\mathrm{gl}(V)$. The first two terms are the Lie algebras of $\mathbb{R}^*$ and $SO(V,g)$. Thus any infinitesimal linear automorphism of (V,g) is the composition of a dilation, a rotation, and a volume preserving map which stretches at possibly different rates along orthogonal axes. Only this last transformation distorts the conformal structure.

Therefore $\mathrm{sym}_0(V,g)$ is naturally identified with the tangent space to $\mathrm{Conf}(V)$ at $[g]$.

A smooth family of metrics g_t with $g_0 = g$ determines a path in $\mathrm{Conf}(V)$ and thus a tangent vector $[\dot{g}_0] \in T_g \,\mathrm{Conf}(V)$. To express $[\dot{g}_0]$ as an element of $\mathrm{sym}_0(V,g)$, let

$$A = \frac{1}{2}g_0^{-1} \circ \dot{g}_0 \;\in\; \mathrm{gl}(V). \tag{A.1}$$

(Here $\dot{g}_0 \in \mathrm{Hom}(V,V^*)$). We claim

$$A - \frac{1}{n}\mathrm{tr}(A)I \;=\; [\dot{g}_0] \in \mathrm{sym}_0(V,g); \tag{A.2}$$

that is, the tangent vector to this path is the trace zero part of A. Indeed,

$$\frac{d}{dt}(I + tA)^* \circ g \circ (I + tA)\Big|_{t=0} \;=\; A^*g + gA$$
$$= \frac{1}{2}(\dot{g}_0^*(g_0^{-1})^*g + gg_0^{-1}\dot{g}_0) \;=\; \dot{g}_0,$$

using the fact that metrics and their derivatives are invariant under $*$. Thus $A \in \mathrm{gl}(V)$ infinitesimally moves g by $\dot{g}_0$, so the projection of A to $\mathrm{sym}_0(V)$ represents $[\dot{g}_0]$.

For $A \in \mathrm{sym}_0(V,g) = T_g \,\mathrm{Conf}(V)$, let

$$\|A\| = \sup_{v \neq 0} \frac{|Av|_g}{|v|_g}. \tag{A.3}$$

This norm is the infinitesimal form of the Teichmüller metric $d(g,h)$. It determines a Finsler structure on $\mathrm{Conf}(V)$ (only in dimension two is this structure Riemannian). For any smooth path g_t, we have

$$d(g_0, g_1) \leq \int_0^1 \| [\dot{g}_t] \| \, dt. \tag{A.4}$$

Here $\| [\dot{g}_t] \|$ is measured using the norm on $T_{g_t} \text{Conf}(V)$. (This formula follows easily from the fact that $K(I+\epsilon A) = 1+n\epsilon\|A\|+O(\epsilon^2)$.)

Any two metrics are joined by a *geodesic* such that equality holds in the expression above; however for $n \geq 3$ these geodesics are generally not unique, even locally.

A.2   Maps and vector fields

A *Riemannian metric* on a smooth n-manifold M is a smoothly varying metric $g(x)$ on the tangent spaces $T_x M$. A *conformal structure* on M is the choice of a Riemannian metric up scale; that is, $g(x)$ and $\alpha(x)g(x)$ determine the same conformal structure, for any smooth function $\alpha(x) > 0$.

Let $f : (M, g) \to (N, h)$ be a homeomorphism between Riemannian n-manifolds.

Definition.   The mapping f is K-*quasiconformal* if the distributional first derivatives of f are locally in L^n, and the differential

$$Df(x) : (T_x M, g(x)) \to (T_{f(x)} N, h(f(x)))$$

satisfies $K(Df(x)) \leq K$ almost everywhere. (If M and N are Riemann surfaces it is also conventional to require f to be orientation-preserving.)

Similarly, f is an L-*quasi-isometry* if $L(Df(x)) \leq L$ almost everywhere (in this case the derivatives are in L^∞).

The simplest conformal manifold is $\mathbb{R}^n$ with the Euclidean metric $|x|^2 = \sum x_i^2$. If f is a homeomorphism between domains in $\mathbb{R}^n$, with the entries in its Jacobian matrix

$$Df = \left[\frac{\partial f_i}{\partial x_j} \right]$$

locally in $L^n(\mathbb{R}^n)$, then f is K-quasiconformal if and only if the eigenvalues of the matrix

$$|\det Df(x)|^{-2/n} Df(x)^* Df(x)$$

lie in the interval $[K^{-2/n}, K^{2/n}]$ for a.e. x.

Next we wish to describe vector fields which generate quasiconformal flows. Let $v : M \to TM$ be a continuous vector field on (M, g). Let

$$Qv = \frac{1}{2} g^{-1} \mathcal{L}_v(g),$$

where $\mathcal{L}_v(g)$ denotes the Lie derivative of the metric g. The *conformal strain* Sv is the distributional section of $\text{sym}_0(TM, g)$ defined by

$$Sv = Qv - \frac{1}{n} \text{tr}(Qv)I, \tag{A.5}$$

where I denotes the identity map; that is, Sv is the traceless part of Qv.

The conformal strain is an infinitesimal linear map of the tangent space to itself, which transports the metric $g(x)$ to the pullback of g under the infinitesimal diffeomorphism generated by v, up to rescaling so the volume element is preserved. In other words,

$$Sv(x) = [\mathcal{L}_v(g)(x)] \in T_{g(x)} \text{Conf}(T_x M).$$

The formula for the strain comes from (A.1) and (A.2).

The conformal strain depends only on the conformal class of g. We can check this directly: for any smooth function $\alpha > 0$,

$$(\alpha g)^{-1} \mathcal{L}_v(\alpha g) = \alpha^{-1} g^{-1}((v\alpha)g + \alpha \mathcal{L}_v(g)) = \alpha^{-1}(v\alpha)I + g^{-1}\mathcal{L}_v(g);$$

removing the trace and dividing by 2, we find the strain relative to αg is the same as for g.

Definition. A vector field v is *k-quasiconformal* if the strain distribution Sv is in L^∞, and if

$$\|Sv\|_\infty = \text{ess. sup}_M \|Sv(x)\| \leq k.$$

Here $\|Sv(x)\|$ is the operator norm determined by $g(x)$ using (A.3).

On a domain in $\mathbb{R}^n$, we can write $v(x) = \sum v_i(x) \, \partial/\partial x_i$; then

$$Sv = \sum_{i,j} (Sv)_{ij} \, dx^i \otimes \frac{\partial}{\partial x_j}$$

where

$$(Sv)_{ij} = \frac{1}{2} \left(\frac{\partial v_i}{\partial x_j} + \frac{\partial v_j}{\partial x_i} \right) - \frac{1}{n} \sum_k \frac{\partial v_k}{\partial x_k}.$$

A Riemannian manifold (M, g) is *conformally flat* if it is locally conformally equivalent to $\mathbb{R}^n$. Riemann surfaces, spheres, and spaces of constant curvature are all conformally flat. Stereographic projection gives a conformal model of the sphere S^n as $\mathbb{R}^n \cup \infty$, with the inversion $R(x) = x/|x|^2$ providing a chart in a neighborhood of infinity. In the sequel we will only be interested in conformally flat manifolds, so the local picture of domains in $\mathbb{R}^n$ will suffice.

Riemann surfaces and Beltrami differentials. On a Riemann surface X, there is a natural sense in which $S = \bar{\partial}$. This is not too surprising, since S annihilates conformal vector fields and $\bar{\partial}$ annihilates holomorphic vector fields.

To make the equation $S = \bar{\partial}$ precise, recall there is a natural splitting $TX \otimes_{\mathbb{R}} \mathbb{C} = T'X \oplus T''X$, where $T'X$ is the holomorphic tangent bundle of X, and $T''X$ is its complex conjugate. For a section $\hat{v} : X \to T'X$, $\bar{\partial}\hat{v}$ is a Beltrami differential, that is a section of $\mathrm{Hom}(T''X, T'X)$.

The composition of $TX \hookrightarrow TX \otimes \mathbb{C}$ with projection to $T'X$ gives a natural isomorphism $TX \cong T'X$. Thus a real vector field v determines a complex vector field $\hat{v} : X \to T'X$, which is characterized by $v = 2\,\mathrm{Re}(\hat{v})$. There is a similar isomorphism

$$\mathrm{sym}_0(TX) \cong \mathrm{Hom}(T''X, T'X) \subset \mathrm{Hom}(TX, TX) \otimes \mathbb{C},$$

also characterized by $s = 2\,\mathrm{Re}(\hat{s})$.

We claim these isomorphisms send the operator S to the operator $\bar{\partial}$; that is,

$$v = 2\,\mathrm{Re}(\hat{v}) \implies Sv = 2\,\mathrm{Re}(\bar{\partial}\hat{v}).$$

This is easily verified in terms of a local coordinate $z = x + iy$. If

$$v = s\frac{\partial}{\partial x} + t\frac{\partial}{\partial y}, \quad \text{then} \quad \hat{v} = (s + it)\frac{\partial}{\partial z},$$

and

$$\bar{\partial}\hat{v} = \frac{\partial}{\partial \bar{z}}(s + it)\frac{d\bar{z}}{dz} = \frac{1}{2}(s_x - t_y + i(s_y + t_x))\frac{d\bar{z}}{dz},$$

where the subscripts denote differentiation. With respect to the basis $\{\partial/\partial x, \partial/\partial y\}$ we have

$$\frac{d\bar{z}}{dz} = (dx - idy) \otimes \frac{1}{2}\left(\frac{\partial}{\partial x} - i\frac{\partial}{\partial y}\right) = \frac{1}{2}\begin{pmatrix} 1 & -i \\ -i & -1 \end{pmatrix},$$

and thus

$$2\operatorname{Re}(\bar{\partial}\hat{v}) = \frac{1}{2}\begin{pmatrix} s_x - t_y & s_y + t_x \\ s_y + t_x & t_y - s_x \end{pmatrix} = Sv.$$

The operator norm defined on $\operatorname{sym}_0(TX)$ by (A.3) corresponds to the usual norm on Beltrami differentials, $\|\mu(z)d\bar{z}/dz\| = |\mu(z)|$.

In the sequel we will treat v and $\hat{v}$ interchangeably, as well as Sv and $\bar{\partial}v$, implicitly making use of the natural isomorphisms just discussed.

For any point x on a Riemann surface X, the space $\operatorname{Conf}(T_x X)$ with its Teichmüller metric is isometric to a hyperbolic plane of constant curvature -4. There is a natural identification between $\operatorname{Conf}(T_x X)$ and the unit ball in $\operatorname{sym}_0(T_x X)$, coming from the Poincaré model for the hyperbolic plane. A smooth map $f : X \to Y$ between Riemann surfaces determines a section $\mu = [f^*g]$ of the bundle of hyperbolic planes $\operatorname{Conf}(TX)$, where g is any metric representing the conformal structure on Y. With respect to the natural identification between $\operatorname{sym}_0(T_x X)$ and the Beltrami differentials at x, the section μ is the same as the *complex dilatation* of f, given in local coordinates by

$$\mu = \frac{f_{\bar{z}}}{f_z}\frac{d\bar{z}}{dz}.$$

Conformal vector fields. A continuous vector field w is *conformal* if $Sw = 0$ as a distribution. By ellipticity considerations, a conformal vector field is smooth. Indeed, on $\mathbb{R}^n$ the condition $Sw = 0$ is equivalent to the system of equations

$$\frac{\partial w_i}{\partial x_j} = -\frac{\partial w_j}{\partial x_i} \quad (i \neq j),$$

$$\frac{\partial w_i}{\partial x_i} = \frac{1}{n}\sum_k \frac{\partial w_k}{\partial x_k}.$$

Thus for $i \neq j$ we have

$$\frac{\partial^2 w_i}{\partial x_j^2} = -\frac{\partial^2 w_j}{\partial x_i \partial x_j} = -\frac{1}{n}\frac{\partial}{\partial x_i}\sum_k \frac{\partial w_k}{\partial x_k} = -\frac{\partial^2 w_i}{\partial x_i^2},$$

which implies

$$(n-1)\frac{\partial^2 w_i}{\partial x_i^2} + \sum_{i\neq j}\frac{\partial^2 w_i}{\partial x_j^2} = 0.$$

But up to an affine change of coordinates this is just Laplace's equation, so any weak solution is automatically smooth.

The global conformal vector fields on the n-sphere form a finite-dimensional space (isomorphic to the Lie algebra $so(n+1,1)$). Regarding S^n as $\mathbb{R}^n \cup \{\infty\}$, $n > 1$, the conformal vector fields are those of the form

$$w(x) \;\; = \;\; w_0 + Ax + 2\langle w_\infty, x\rangle x - |x|^2 w_\infty.$$

Here w_0 and w_∞ are vectors in $\mathbb{R}^n$, and $A : \mathbb{R}^n \to \mathbb{R}^n$ is a linear map whose symmetric trace-free part vanishes. Regarded as vector fields on the sphere, the first two terms vanish at infinity; the last two terms give a parabolic vector field, vanishing at $x = 0$, but nonzero at infinity when $w_\infty \neq 0$. When $n = 2$ we may identify S^2 with $\widehat{\mathbb{C}}$; then the conformal vector fields are given by

$$w(z) = (w_0 + az + w_\infty z^2)\frac{\partial}{\partial z},$$

where w_0, a and w_∞ are arbitrary complex numbers.

Further properties of quasiconformal maps.

1. A K-quasiconformal mapping is absolutely continuous with Radon-Nikodym derivative equal to $|\det Df|$ almost everywhere [Vai, Thm. 33.3]. For any ball B,

$$\int_B \left|\frac{\partial f_i}{\partial x_j}\right|^n \;\; \leq \;\; K\int_B |\det Df| \;\; = \;\; \mathrm{vol}(fB);$$

   this is why one expects derivatives locally in L^n.

2. A homeomorphism f between conformally flat manifolds is K-quasiconformal if and only if any path family Γ of finite modulus satisfies

$$\frac{1}{K}\,\mathrm{mod}(\Gamma) \leq \mathrm{mod}(f(\Gamma)) \leq K\,\mathrm{mod}(\Gamma).$$

See [Vai, Thm. 32.3].

In dimension $n \geq 3$ there are many possible alternative measures of the dilatation of a quasiconformal map; see [Vai], [Res] and [Ah4] for examples. The direct connection with the modulus of a path family makes the present definition especially convenient. For example, it facilitates the proof of:

3. *A homeomorphism $f : (M, g) \rightarrow (N, h)$ which is a uniform limit of K-quasiconformal homeomorphisms is itself K-quasiconformal.* For domains in $\mathbb{R}^n$ this is proved in [Vai, Cor. 37.3]. The generalization to Riemannian manifolds is immediate, since on a small scale any continuous Riemannian metric is almost isometric to a flat metric.

4. A K-quasiconformal map $f : B^n \rightarrow B^n$ defined on the open unit ball extends by continuity to a quasiconformal map on $\partial B^n = S^{n-1}$. Moreover f extends by reflection to a K-quasiconformal map on S^n [Vai, Theorem 35.2].

5. In dimension $n = 1$ it is more natural to use a geometric definition of quasiconformality; one obtains the classes of *quasisymmetric* functions, which need not even be absolutely continuous.

6. General references on quasiconformal geometry in higher dimensions include [Res], [Vai], [Vuo1] and [Vuo2].

A.3   BMO and Zygmund class

To establish regularity of quasiconformal vector fields, we begin with a result from real analysis.

Let $f : \mathbb{R}^n \rightarrow \mathbb{R}$ be continuous, and let $\nabla f = \langle \partial f / \partial x_i \rangle$ be the distributional gradient of f. It is easy to see that f is Lipschitz when ∇f is in L^∞. In this section we will show that f is almost Lipschitz when ∇f is almost in L^∞.

Theorem A.2 *Let f be a compactly supported continuous function on $\mathbb{R}^n$. If ∇f has bounded mean oscillation, then f is in the Zygmund class, and*

$$\|f\|_Z \ \leq \ C_n \|\nabla f\|_{BMO}.$$

(Here and in the sequel, C_n denotes a constant that depends only on n.)

Definitions. The *Zygmund norm* of f is

$$\|f\|_Z \;=\; \sup_{x,y\in\mathbb{R}^n,\,y\neq0} \frac{|f(x+y)+f(x-y)-2f(x)|}{|y|},$$

where $|y|$ denotes the Euclidean norm of y. A function is in the *Zygmund class* if $\|f\|_Z < \infty$.

The *BMO norm* of f is

$$\|f\|_{BMO} \;=\; \sup \frac{1}{|B|}\int_B |f(x)-f_B|\,dx,$$

where the supremum is over all round balls $B \subset \mathbb{R}^n$, $|B|$ denotes the volume of B and f_B denotes the mean value of f on B (by convention, $\|f\|_{BMO} = \infty$ if f is not locally integrable). If $\|f\|_{BMO} < \infty$, f is of *bounded mean oscillation*.

Each "norm" is actually a pseudo-norm; $\|f\|_{BMO} = 0$ if f is constant, and $\|f\|_Z = 0$ if the gradient of f is constant. BMO occurs naturally in connection with quasiconformal mappings, because $\|f\|_{BMO}$ is scale-invariant. For example, $\log|\det D\phi|$ is in BMO for any quasiconformal map $\phi : \mathbb{R}^n \to \mathbb{R}^n$, and if a homeomorphism $\phi : \mathbb{R}^n \to \mathbb{R}^n$ satisfies $\|f\circ\phi\|_{BMO} \le C\|f\|_{BMO}$ for every f in BMO, then ϕ is quasiconformal; see [Rei1].

We say f has *modulus of continuity* $\omega(r)$ if

$$|f(x)-f(y)| \le \omega(|x-y|).$$

For example, one may take $\omega(r) = O(r)$ if f is Lipschitz. A Zygmund function is almost Lipschitz, in the following sense:

Proposition A.3 *If $f : \mathbb{R}^n \to \mathbb{R}$ is a bounded function in the Zygmund class, then f has modulus of continuity*

$$\omega(r) = Mr\left(1+\overset{+}{\log}\frac{1}{r}\right),$$

where $M = C(\|f\|_Z + \|f\|_\infty)$ for a universal constant C.

Here $\overset{+}{\log} x = \max(0, \log x)$.

Proof. Given any linear segment $I = [x, y] \subset \mathbb{R}^n$, let the *slope* of f over I be the ratio $(f(x) - f(y))/|x - y|$. The Zygmund condition says if we cut I into two subintervals I_1 and I_2 of equal length, then the slope over I_1 or I_2 differs a bounded amount from the slope over I. After subdiving n times we reach segments of length $2^{-n}|I|$. The slope over a segment of length one or more is bounded by $2\|f\|_\infty$, so the slope over a segment of length greater than 2^{-n} is at most $O(\|f\|_\infty + n\|f\|_Z)$.

∎

Similarly, $\|f\|_{BMO} < \infty$ implies f is almost bounded; for instance f is locally in L^p for all $p < \infty$. More precisely we have from [JN]:

Theorem A.4 (John-Nirenberg) *For any ball B, and $1 \le p < \infty$,*

$$\left(\frac{1}{|B|} \int_B |f(x) - f_B|^p dx \right)^{1/p} \le C_{p,n} \|f\|_{BMO}.$$

Proof of Theorem A.2. Note that the Theorem is immediate on $\mathbb{R}^1$.

The function $f(x)$ can be recovered from its gradient by convolving with the vector-valued kernel

$$K(x) = \frac{1}{\omega_{n-1}} \frac{x}{|x|^n}$$

(where $x = (x_1, \dots, x_n)$ and ω_{n-1} is the volume of the unit $(n-1)$-sphere in $\mathbb{R}^n$). That is,

$$f(x) = (\nabla f * K)(x) = \int_{\mathbb{R}^n} \langle \nabla f(y), K(x - y) \rangle \, dy.$$

(See [Stein2, V.2.3]). Note that $K(x)$ is homogeneous of degree $1 - n$.

To prove the Theorem, it suffices to show

$$|f(e_1) + f(-e_1) - 2f(0)| \le C_n \|\nabla f\|_{BMO}$$

(where $e_1 = (1, 0, \dots, 0)$), since any quotient appearing in the Zygmund norm can be translated into the form above by precomposing f with a suitable similarity, and this operation leaves the ratio $\|f\|_Z / \|\nabla f\|_{BMO}$ unchanged.

Equivalently, we must bound

$$\left| \int \langle \nabla f(x), L(x) \rangle \, dx \right|$$

in terms of $\|\nabla f\|_{BMO}$, where

$$L(x) \;=\; K(e_1 - x) + K(-e_1 - x) - 2K(-x).$$

To this end, let $\mathbb{R}^n = A_0 \cup A_1 \cup A_2 \cup \ldots$, where A_0 is the ball of radius 2 centered at the origin, and $A_k = \{x \ : \ 2^k < |x| \le 2^{k+1}\}$ for $k > 0$. Let M_k denote the mean of ∇f over A_k. Since $\int_{\mathbb{R}^n} L = 0$, replacing ∇f by $\nabla f - M_0$ leaves $\int \langle \nabla f, L \rangle$ unchanged, so we may assume $M_0 = 0$. From the definition of BMO we find by induction

$$|M_k| \;=\; O(k \|\nabla f\|_{BMO});$$

moreover for $1 \le p < \infty$, Theorem A.4 implies the L^p norm of $\nabla f - M_k$ restricted to A_k satisfies

$$\|(\nabla f - M_k)|A_k\|_p \;\le\; O(|A_k|^{1/p} \|\nabla f\|_{BMO}) \qquad (\text{A.6})$$

(where the implied constants depend only on p and n).

We now bound $|\int \langle \nabla f, L \rangle|$ by breaking the integral into three pieces.

First, on A_0, $L(x)$ has three singularities of type $1/|x|^{n-1}$, so $L|A_0$ is in L^q for some q slightly greater than 1. Since $M_0 = 0$ we may apply (A.6) plus Hölder's inequality to conclude

$$\left| \int_{A_0} \langle \nabla f, L \rangle \right| \;\le\; \|\nabla f|A_0\|_p \, \|L|A_0\|_q \;=\; O(\|\nabla f\|_{BMO}).$$

Similarly, for $|x| > 2$, $L(x) = O(1/|x|^{n+1})$; so for $k > 0$,

$$\|L|A_k\|_q = O(2^{-k(n+1)} |A_k|^{1/q}).$$

Choosing $p = q = 2$, Hölder's inequality implies

$$\sum_{k>0} \left| \int_{A_k} \langle \nabla f - M_k, L \rangle \right| \;\le\; O\left(\sum 2^{-k(n+1)} |A_k| \, \|\nabla f\|_{BMO} \right)$$

$$= \; O(\|\nabla f\|_{BMO})$$

since $|A_k| = O(2^{nk})$.

Finally, the L^1-norm of $L|A_k$ is $O(2^{-k})$, so

$$\sum_{k>0} \left| \int_{A_k} \langle M_k, L \rangle \right| = O\left(\sum 2^{-k} k \|\nabla f\|_{BMO} \right) = O(\|\nabla f\|_{BMO})$$

as well. The sum of these three estimates proves the Theorem.

∎

Hölder continuity. Under the weaker assumption that ∇f is in L^p, one may still conclude that f is Hölder continuous. Thus the Theorem above can be considered as a limiting case of the more traditional Sobolev estimate:

Theorem A.5 *If f is a compactly supported continuous function on $\mathbb{R}^n$ with distributional derivatives in L^p, $n < p < \infty$, then f is Hölder continuous of exponent $\alpha = 1 - n/p$, and*

$$|f(x) - f(y)| \leq C_{p,n} \|\nabla f\|_p |x - y|^\alpha.$$

Proof. After changing coordinates by a Euclidean similarity, we can reduce to the case $x = 0$ and $y = e_1$. Thus it suffices to show

$$|f(0) - f(e_1)| \leq C_{p,n} \|\nabla f\|_p.$$

Equivalently, we must bound

$$\left| \int \langle \nabla f(x), L(x) \rangle \, dx \right|$$

where

$$L(x) = K(-x) - K(e_1 - x)$$

and, as before, $K(x) = (1/\omega_{n-1})(x/|x|^n)$. Near its two singularities, $|L(x)|$ behaves like $1/|x|^{n-1}$, while for x large it behaves like $|x|^n$. Thus $L(x)$ is in L^q for $1 < q < n/(n-1)$, and the Theorem follows by Hölder's inequality.

∎

See also [SZ, Theorem 2(b)], [Zie, Thm. 2.4.4].

A.4   Compactness and modulus of continuity

In this section we show a quasiconformal vector field v has an $r \log(1/r)$ modulus of continuity. This estimate is essential for unique integrability. We also show that when the strain Sv is small (in L^∞ or L^p), then v is uniformly close to a *conformal* vector field.

The key to these results is the fact that a smooth, compactly supported vector field v on $\mathbb{R}^n$ can be recovered from its strain Sv using a Calderón-Zygmund operator. The general theory of such operators shows Sv controls ∇v, and the latter controls v via the Sobolev-type bounds of the preceding section.

Definitions. Suppose the kernel $K : (\mathbb{R}^n - \{0\}) \to \mathbb{R}$ is homogeneous of degree $-n$, and its restriction to the unit sphere is smooth and of mean zero. Then the transformation

$$T(f)(x) \;=\; (K * f)(x) \;=\; \int K(y)f(x-y)dy$$

is a *Calderón-Zygmund* operator on $\mathbb{R}^n$.

These operators arise naturally in the context of conformal geometry. For example, the condition that K is homogeneous of degree $-n$ says exactly that T commutes with real dilations: $(Tf)(\lambda x) = T(f(\lambda x))$.

The integral above is intended in the sense of the principal value

$$\lim_{r \to 0} \int_{\mathbb{R}^n - B(0,r)} K(y)f(x-y)\, dy.$$

It then makes sense for all compactly supported smooth f, and one can consider its properties as an operator between various function spaces. For example, T is a bounded operator on L^p for $1 < p < \infty$. For $p = \infty$ we have the following result:

Theorem A.6 *Let T be a Calderón-Zygmund operator, and f a bounded measurable function with compact support. Then Tf has bounded mean oscillation and*

$$\|Tf\|_{BMO} \;\leq\; C(T)\|f\|_\infty.$$

See [Stein1, Thm. 4], which includes a survey of Calderón-Zygmund theory.

Example. Let $v = v(z)\partial/\partial z$ be a smooth, compactly supported complex vector field on $\mathbb{C}$. Then

$$\frac{\partial v}{\partial z}(\zeta) = -\frac{1}{\pi} \int_{\mathbb{C}} \frac{1}{z^2} \frac{\partial v}{\partial \bar{z}}(\zeta - z) |dz|^2.$$

The transformation $v_{\bar{z}} \mapsto T(v_{\bar{z}}) = v_z$ is a Calderón-Zygmund operator with kernel $K(z) = -\pi/z^2$. If v is quasiconformal, then $\|\bar{\partial}v\|_\infty < \infty$, so by Theorem A.6 the derivatives of v are in BMO.

To state a generalization of this classical formula to $\mathbb{R}^n$, for $v = \sum v_i \, \partial/\partial x_i$ let

$$(Av)_{ij} = \frac{1}{2}\left(\frac{\partial v_i}{\partial x_j} - \frac{\partial v_j}{\partial x_i}\right) + \frac{1}{n}\sum_k \frac{\partial v_k}{\partial x_k},$$

so that $(\nabla v_i)_j = (Av)_{ij} + (Sv)_{ij}$. Then we have from [Ah1]:

Theorem A.7 (Ahlfors) *There is a matrix-valued Calderón-Zygmund kernel $K_{ij}(x)$ such that for any smooth, compactly supported vector field v on $\mathbb{R}^n$,*

$$(Av)_{ij} = K_{ij} * Sv.$$

Corollary A.8 *Let $v = \langle v_i(x) \rangle$ be a compactly supported quasiconformal vector field on $\mathbb{R}^n$. Then the distributional derivatives of v have bounded mean oscillation, and*

$$\left\|\frac{\partial v_i}{\partial x_j}\right\|_{BMO} \leq C_n \|Sv\|_\infty.$$

Proof. Combining the last two Theorems, we have $\|Av\|_{BMO} \leq C_n\|Sv\|_\infty$. Since $\|Sv\|_{BMO} \leq 2\|Sv\|_\infty$, the BMO norm of $\nabla v = Av + Sv$ is bounded by $(C_n + 2)\|Sv\|_\infty$.

∎

Applying Theorem A.2 and Proposition A.3 we have:

Corollary A.9 *The components of v are in Zygmund class, with $\|v_i\|_Z \leq C_n\|Sv\|_\infty$, and v has modulus of continuity $Mr(1+\overset{+}{\log}(1/r))$, where $M = C_n(\|Sv\|_\infty + \|v\|_\infty)$.*

Since the sphere is compact, quasiconformal vector fields on S^n satisfy bounds similar to those for compactly supported vector fields on $\mathbb{R}^n$. However we must take into account the fact that the sphere carries a finite-dimensional family of conformal vector fields.

Theorem A.10 *Given a quasiconformal vector field v on S^n, there is a conformal vector field w such that*

$$\|v - w\|_\infty \leq C_n \|Sv\|_\infty.$$

Moreover $(v-w)$ has modulus of continuity $\omega(r) = Mr(1 + \overset{+}{\log}(1/r))$, where $M = C_n \|Sv\|_\infty$.

Here the size of $(v-w)$ and its modulus of continuity are measured using the spherical metric and parallel transport. Equivalently, one can measure the size of v using the Euclidean geometry of a fixed pair of conformally flat charts.

Proof. Let $\tilde{v}$ denote the image of v in the space of quasiconformal vector fields modulo conformal vector fields, and let

$$\|\tilde{v}\|_\infty = \inf\{\|v - w\|_\infty : Sw = 0\}$$

be the quotient L^∞-norm. We will first show there exists a constant C_n such that

$$\|\tilde{v}\|_\infty \leq C_n \|Sv\|_\infty.$$

The proof is by contradiction. If no such C_n exists, then we can find a sequence v_k such that

$$\|\widetilde{v_k}\|_\infty = 1 \quad \text{while} \quad \|Sv_k\|_\infty \to 0.$$

We can also assume $\|v_k\|_\infty = 1$.

Regarding S^n as $\mathbb{R}^n \cup \{\infty\}$, choose a smooth function f with compact support on $\mathbb{R}^n$ such that $f(x) = 1$ when $|x| \leq 1$. The strain $S(fv_k)$ is the sum of fSv_k and terms involving v_k and derivatives of f. Thus

$$\|S(fv_k)\|_\infty = O(\|Sv_k\|_\infty + \|v_k\|_\infty) = O(1);$$

that is, it is bounded independent of k. But fv_k is compactly supported on $\mathbb{R}^n$, so by Corollary A.9, $\|fv_k\|_Z = O(1)$ and v_k has a uniform modulus of continuity on $\{x : |x| \le 1\}$. Rotating the sphere, we may use the same argument to conclude that v_k has a uniform modulus of continuity on all of S^n. Since $\|v_k\|_\infty = 1$, the Arzela-Ascoli theorem provides a subsequence such that $v_k \to v_\infty$ uniformly. But $\|Sv_k\|_\infty \to 0$, so $Sv_\infty = 0$ and therefore v_∞ is conformal. Since $\|v_k - v_\infty\|_\infty \to 0$, we have $\|\widetilde{v_k}\|_\infty \to 0$, a contradiction.

Therefore $\|\tilde{v}\|_\infty = O(\|Sv\|_\infty)$. So after correcting v by a conformal vector field w, we have $\|v\|_\infty \le C_n \|Sv\|_\infty$ as desired. Moreover $\|S(fv)\|_\infty = O(\|Sv\|_\infty)$, so the modulus of continuity of the corrected vector field v is also controlled by $\|Sv\|_\infty$.

■

Corollary A.11 *For fixed $k \ge 1$, the space of k-quasiconformal vector fields on S^n, modulo conformal vector fields, is compact in the uniform topology.*

Proof. By the preceding Theorem, any sequence $\tilde{v}_i$ has a uniformly convergent subsequence. By general properties of distributional derivatives,

$$\|S(\lim \tilde{v}_i)\|_\infty \le \liminf \|S\tilde{v}_i\|_\infty \le k$$

so the limit is still k-quasiconformal.

■

Corollary A.12 *A quasiconformal vector field on S^n has modulus of continuity $\omega(r) = Mr(1 + \overset{+}{\log}(1/r))$, where $M = C_n(\|v\|_\infty + \|Sv\|_\infty)$.*

Proof. The strain $\|Sv\|_\infty$ bounds the modulus of continuity of $(v - w)$ and the size of $\|v - w\|_\infty$, for some conformal vector field w. From $\|v\|_\infty$ we obtain a bound on $\|w\|_\infty$, which in turn controls the Lipschitz constant of w. Adding w back in, we obtain control of the modulus of continuity of v.

■

L^p bounds on strain and Hölder continuity. Additional control of v can sometimes be obtained from the L^p-norm

$$\|Sv\|_p \;=\; \left(\int_{S^n} \|Sv(x)\|^p \, dV(x)\right)^{1/p},$$

where $dV(x)$ is the spherical volume element. For example, this norm takes into account the measure of the support of Sv, while $\|Sv\|_\infty$ does not. Parallel to Theorem A.10 we have:

Theorem A.13 *Let v be a continuous vector field on S^n with strain Sv in L^p, $n < p < \infty$. Then there is a conformal vector field w such that*

$$\|v - w\|_\infty \;\leq\; C_{p,n}\|Sv\|_p.$$

Moreover $(v - w)$ has a Hölder modulus of continuity $\omega(r) = Mr^\alpha$, where $\alpha = 1 - n/p$ and $M = C_{p,n}\|Sv\|_p$.

Proof. First let v be a smooth compactly supported vector field on $\mathbb{R}^n$. By Ahlfors' Theorem A.7, $(\nabla v - Sv) = T(Sv)$ for a (matrix-valued) Calderón-Zygmund operator T. Since T is bounded on L^p [Stein2, §II.4.2], we have $\|\nabla v\|_p = O(\|Sv\|_p)$. Then $\|\nabla v\|_p$ controls the modulus of Hölder continuity of v by the Sobolev-type Theorem A.5.

Now let v be a vector field on $S^n = \mathbb{R}^n \cup \{\infty\}$. As in the proof of Theorem A.10, by applying the bounds on $\mathbb{R}^n$ to truncated vector fields fv, we may bound the modulus of continuity of v in terms of $\|v\|_\infty$ and $\|Sv\|_p$. The same compactness argument as before then establishes

$$\inf_{Sw=0} \|v - w\|_\infty = O(\|Sv\|_p).$$

It follows that $\|Sv\|_p$ also controls the modulus of continuity of $(v - w)$. ■

A.5   Unique integrability

A vector field is *uniquely integrable* if it generates a well-defined flow. The following criterion for unique integrability is classical:

Proposition A.14 (Osgood) *A continuous, compactly supported vector field $v_t(x)$ on $\mathbb{R}^n$ is uniquely integrable if*

$$|v_t(x) - v_t(y)| \leq \omega(|x - y|)$$

for all t, and $\displaystyle\int_0^1 \frac{dr}{\omega(r)} = \infty$.

Heuristically, if ϕ_t^1 and ϕ_t^2 are two flows generated by v_t, then $\Delta(t) = \sup |\phi_t^1(x) - \phi_t^2(x)|$ satisfies $|d\Delta/dt| \leq \omega(\Delta)$, which gives

$$\infty = \int_{t=0}^{t=1} \frac{|d\Delta|}{\omega(\Delta)} \leq \int_0^1 dt = 1$$

if Δ is not identically zero. See [Hil, p.59].

Example. A field of vectors tangent to the horizontal translates of the graph of $y = f(x) = x^3$ in the plane is *not* uniquely integrable (see Figure A.1). The x-axis and the graph of f are both integral curves through $(0,0)$.

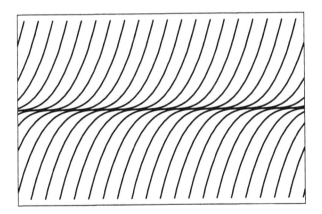

Figure A.1. A non-uniquely integrable line field.

Since $\int_0^1 dr/|r \log r| = \infty$, Osgood's criterion and Corollary A.12 imply:

Corollary A.15 *A continuous quasiconformal vector field*

$$v : S^n \times [0,1] \to TS^n$$

with $\|Sv_t\|_\infty \leq k$ for all t is uniquely integrable.

We are now in a position to prove a quasiconformal vector field on S^n generates a unique quasiconformal flow.

Proof of Theorem A.1. Existence of a flow ϕ_t integrating v_t is classical, by compactness of $S^n \times [0,1]$ and continuity of v_t. Uniqueness of ϕ_t is given by the preceding Corollary.

Next we verify quasiconformality of ϕ_t under the assumption that v_t is smooth. Let g_0 be the standard metric on S^n, and let $g_t = \phi_t^*(g_0)$. Then for each $x \in S^n$, we obtain a family of metrics $g_t(x)$ and hence a path in $\mathrm{Conf}(T_x S^n)$. As a section of $\mathrm{sym}_0(TS^n, g_t)$, we have

$$[dg_t/dt] = [\phi_t^* \mathcal{L}_{v_t}(g_0)] = \phi_t^*(Sv_t),$$

using (A.1), (A.2) and (A.5). The Teichmüller metric on $\mathrm{Conf}(T_x S^n)$ is invariant under linear automorphisms of the tangent space, so $\| [dg_t(x)/dt] \| = \|Sv_t(\phi_t(x))\| \leq k$. Now

$$K(D\phi_t) = \exp(nd(g_0(x), g_t(x))),$$

where d is the Teichmüller metric, and by (A.4) we have

$$d(g_0(x), g_t(x)) \leq kt.$$

Thus ϕ_t is K_t-quasiconformal with $K_t = \exp(nkt)$ as stated.

To reduce the general case to the smooth case, let $\alpha_i = \alpha_i(g)dg$ be a family of smooth probability measures on the isometry group G of S^n, converging weakly to the point mass at the identity. Then we obtain a sequence of smooth vector fields v_t^i tending uniformly to v_t by setting

$$v_t^i = \alpha_i * v_t = \int_G g_*(v_t) \, \alpha_i(g) \, dg.$$

Since G acts conformally and the strain operator is linear, we have $\|Sv_t^i\|_\infty \le \|Sv_t\|_\infty \le k$ for all i. Integrating v_t^i, we obtain a flow ϕ_t^i on S^n such that ϕ_t^i is a K_t-quasiconformal homeomorphism. By unique integrability, $\phi_t^i \to \phi_t$ uniformly as $i \to \infty$. But a uniform limit of K_t-quasiconformal mappings is K_t-quasiconformal, establishing the Theorem for v_t.

∎

Appendix B

Visual extension

In this Appendix we study natural extensions of vector fields and flows from the sphere at infinity to hyperbolic space.

For functions, vector fields, and strain fields, there are essentially unique linear operators which provide extensions from S_∞^{n-1} to $\mathbb{H}^n$ and commute with hyperbolic isometries. We consolidate results of Ahlfors, Reimann, and Thurston by showing a quasiconformal vector field v on the sphere extends naturally to a quasi-isometric, volume-preserving and strain-energy minimizing vector field $\text{ex}(v)$ on $\mathbb{H}^n$. A systematic approach to the various extension operators, based on representation theory, follows.

Next we introduce the *visual distortion $Mv(p)$* of a vector field on the sphere, as seen from $p \in \mathbb{H}^n$. By definition $Mv(p)$ is the minimum, over all conformal vector fields w, of the maximum visual length of $(v - w)$ as seen from p. The function $Mv(p)$ is bounded on $\mathbb{H}^n$ if and only if v is quasiconformal. We establish a maximum principle for Mv and other properties that are useful for bounding the visual distortion when v is highly oscillatory.

We then use the visual extension to prolong quasiconformal isotopies on the sphere to quasi-isometric isotopies on hyperbolic space. If the visual distortion is small along a flow line, then the resulting isotopy has a small pointwise quasi-isometry constant. Similarly, if Mv_t tends to zero exponentially fast along a geodesic ray, then the isotopy is $C^{1+\alpha}$-conformal at the endpoint x on the sphere at infinity. These properties are the key to the inflexibility results stated as Theorems 2.11, 2.18, 2.19 and 9.15.

We conclude with an example where the visual extension can be computed explicitly.

B.1 Naturality, continuity and quasiconformality

Definitions. Given a vector bundle $E \to M$ over a smooth manifold M, let $C^\infty(M, E)$ and $\mathcal{D}(M, E)$ denote the spaces of *smooth* and *distributional* sections of E. The first space is equipped with the topology of C^∞ convergence on compact sets. The second space is the completion of the first with respect to the weak topology, in which $\sigma_n \to 0$ if for some smooth measure dV on M, $\int \langle \sigma_n, \phi \rangle \, dV \to 0$ for every compactly supported smooth section ϕ of the dual bundle E^*. For brevity $\mathbb{R}$ and $\mathbb{C}$ will denote trivial bundles.

Consider the Poincaré model for hyperbolic space $\mathbb{H}^n$ as the open unit ball in $\mathbb{R}^n$, bounded by S_∞^{n-1}. We equip S_∞^{n-1} with the usual metric induced from $\mathbb{R}^n$. Let $F(\mathbb{H}^n)$, $V(\mathbb{H}^n)$ and $S(\mathbb{H}^n)$ denote the spaces of smooth functions, vector fields and strain fields on hyperbolic space, and similarly for S_∞^{n-1}. Since isometries of hyperbolic space extend to conformal maps on the sphere, the group $\mathrm{Isom}(\mathbb{H}^n)$ acts on all these spaces, by $g \cdot \omega = g_*(\omega)$. A continuous linear map T between two of these spaces is *natural* if $T(g \cdot \omega) = g \cdot T(\omega)$ for all isometries g (including those that reverse orientation).

We may now state:

Theorem B.1 (Natural maps) *The spaces of natural maps*

$$
\begin{aligned}
F(S_\infty^{n-1}) &\;\to\; F(\mathbb{H}^n), & n \geq 1, \\
V(S_\infty^{n-1}) &\;\to\; V(\mathbb{H}^n), & n \geq 2, & \quad and \\
S(S_\infty^{n-1}) &\;\to\; S(\mathbb{H}^n), & n \geq 3
\end{aligned}
$$

are one-dimensional, as are the spaces of natural maps

$$
\begin{aligned}
F(S_\infty^{n-1}) &\;\to\; V(\mathbb{H}^n), & n \geq 1, \\
F(S_\infty^{n-1}) &\;\to\; S(\mathbb{H}^n), & n \geq 2, & \quad and \\
V(S_\infty^{n-1}) &\;\to\; S(\mathbb{H}^n), & n \geq 2.
\end{aligned}
$$

Each map extends continuously to a map from distributional objects on the sphere to smooth objects on hyperbolic space.

On the other hand, any natural map

$$
\begin{aligned}
V(S_\infty^{n-1}) &\;\to\; F(\mathbb{H}^n), \\
S(S_\infty^{n-1}) &\;\to\; F(\mathbb{H}^n), & \quad or \\
S(S_\infty^{n-1}) &\;\to\; V(\mathbb{H}^n)
\end{aligned}
$$

is zero.

Briefly, there are essentially unique natural maps which preserve or raise the complexity of tensors, but none which lower complexity. The conditions on n rule out the cases where the domain or range is trivial; similar restrictions implicitly apply in the results below. The proof of Theorem B.1 is deferred to the next section.

Next we explicitly describe these natural operators.

Functions. We begin with the map

$$\mathrm{av} : C^\infty(S_\infty^{n-1}, \mathbb{R}) \to C^\infty(\mathbb{H}^n, \mathbb{R}).$$

For $x = 0$ the origin in $\mathbb{R}^n$ (and the center of the sphere), define

$$\mathrm{av}(f)(0) = \frac{1}{\mathrm{vol}(S_\infty^{n-1})} \int_{S_\infty^{n-1}} f(x) \, dV(x),$$

the average of f over S_∞^{n-1}. Since the metric on the sphere is invariant under rotation, $\mathrm{av}(g \cdot f)(0) = \mathrm{av}(f)(0)$ for all $g \in O(n)$. The function $\mathrm{av}(f)$ is extended to all of hyperbolic space by the formula

$$\mathrm{av}(f)(x) = \mathrm{av}(g \cdot f)(0),$$

where $g \in \mathrm{Isom}(\mathbb{H}^n)$ satisfies $g(x) = 0$. This definition is unambiguous because $O(n)$ is the stabilizer of 0 in $\mathrm{Isom}(\mathbb{H}^n)$.

By construction, $\mathrm{av}(f)$ is natural. Intrinsically, $\mathrm{av}(f)(x)$ is the *visual average* of f as seen from the point $x \in \mathbb{H}^n$. That is, if a direction at x is chosen at random (using the hyperbolic metric), then $\mathrm{av}(f)(x)$ is the expected value of f at the endpoint of the corresponding geodesic ray.

Proposition B.2 *The visual average* $\mathrm{av}(f)$ *of any distribution f is a harmonic function in the hyperbolic metric.*

Proof. The map $T : f \mapsto \Delta \, \mathrm{av}(f)$ is a natural map from functions to functions. By Theorem B.1, $T(f)$ is a constant multiple of $\mathrm{av}(f)$. But T sends constants to zero, so $T = 0$.

■

When $n = 2$, the hyperbolic and Euclidean Laplacians are the same, and $\mathrm{av}(f)$ is given by the Poisson kernel.

Vector fields. Now let v be a vector field on the sphere. Since $\mathbb{R}^n$ is a vector space, its tangent bundle is canonically trivial; every tangent vector to the sphere can be translated to the origin. Using this translation, we define

$$\mathrm{av}(v)(0) = \frac{1}{\mathrm{vol}(S_\infty^{n-1})} \int_{S_\infty^{n-1}} v(x)\, dV(x).$$

Clearly acting on v by a rotation of the sphere changes $\mathrm{av}(v)(0)$ by the same rotation. Thus $\mathrm{av}(v)(x)$ can be defined unambiguously by first changing coordinates with a hyperbolic isometry so $x = 0$, using the definition above, then transporting back to x. The result is a natural map

$$\mathrm{av} : C^\infty(S_\infty^{n-1}, \mathrm{T}S_\infty^{n-1}) \to C^\infty(\mathbb{H}^n, \mathrm{T}\mathbb{H}^n).$$

Proposition B.3 *In the Euclidean metric on $\mathbb{H}^n \cup S_\infty^{n-1}$, we have*

$$\sup_{p \in \mathbb{H}^n} |\mathrm{av}(v)(p)| \le 4 \sup_{S_\infty^{n-1}} |v|.$$

Proof. For any hyperbolic isometry g and any $x \in S_\infty^{n-1}$, $|g'(0)| \le 4|g'(x)|$. (This bound is clear when $n = 2$ and $g(z) = (z-a)/(1-\bar{a}z)$, $|a| < 1$, in complex coordinates. The case of $n > 2$ reduces to this case by composing g with an element of $O(n)$ so a 2-plane through 0 and x is preserved.)

Choose an isometry g such that $g(0) = p$. Then by naturality,

$$
\begin{aligned}
|\mathrm{av}(v)(p)| &= |g'(0)| \cdot |\mathrm{av}(g^*v)(0)| \\
&\le |g'(0)| \frac{1}{\mathrm{vol}(S_\infty^{n-1})} \int_{S_\infty^{n-1}} \frac{|v(g(x))|}{|g'(x)|} \, dV(x) \le 4 \sup |v|
\end{aligned}
$$

as claimed.   ■

The averaging map on vector fields can be described intrinsically as follows. First, the exponential map identifies the sphere at infinity with the unit sphere S_x^{n-1} in the tangent space to $x \in \mathbb{H}^n$. Then, a tangent vector on the sphere at infinity gives a tangent vector to $S_x^{n-1} \subset T_x \mathbb{H}^n$, and hence a vector at x. Averaging over S_x^{n-1} yields $\mathrm{av}(v)(x)$.

Strain fields, etc. The same procedure can be applied to tensors of any type; in particular, we obtain a natural map $\mu \mapsto \mathrm{av}(\mu)$ defined on strain fields in this way. Using invariant differential operators on $\mathbb{H}^n$, one can raise the complexity of tensors. The natural maps $f \mapsto \nabla \mathrm{av}(f)$, $f \mapsto S\nabla \mathrm{av}(f)$ and $v \mapsto S \mathrm{av}(v)$ are of this form. Together with $\mathrm{av}(\cdot)$, these maps represent all the nonzero operators appearing in Theorem B.1.

Continuity. For a continuous function f, it is easy to see that f fits together with $\mathrm{av}(f)$ to give a continuous function on $S_\infty^{n-1} \cup \mathbb{H}^n$. For vectors and other tensors, we need to correct by a constant factor to achieve this continuity, because of cancellations due to phase. To this end we define the *extension operator* by

$$\mathrm{ex}(f) = \mathrm{av}(f) \qquad \text{for functions,}$$

$$\mathrm{ex}(v) = \frac{n}{2(n-1)} \mathrm{av}(v) \qquad \text{for vector fields, and}$$

$$\mathrm{ex}(\mu) = \frac{n}{n-2} \mathrm{av}(\mu) \qquad \text{for strain fields.}$$

Theorem B.4 (Continuous extension) *If ω is a continuous function, vector field or strain field on S_∞^{n-1}, then $\omega \cup \mathrm{ex}(\omega)$ is continuous on $S_\infty^{n-1} \cup \mathbb{H}^n$.*

To give the proof, it is convenient to use both the ball model and the upper half-space model for hyperbolic space. We will use coordinates $x = (x_1, \ldots, x_n)$ for the ball model (where $\mathbb{H}^n = \{x : |x| < 1\}$), and coordinates $y = (y_1, \ldots, y_n)$ for the upper half-space model (where $\mathbb{H}^n = \{(y_1, \ldots, y_n) : y_n > 0\}$). In these coordinates, the hyperbolic metric is given by

$$\rho = \frac{2|dx|}{1 - |x|^2} = \frac{|dy|}{y_n}.$$

In the upper half-space model, the boundary of hyperbolic space is $\mathbb{R}^{n-1}_\infty \cup \{\infty\}$, where $\mathbb{R}^{n-1}_\infty$ is the plane $y_n = 0$. To relate the two models, identify $\mathbb{R}^{n-1}_\infty$ with the plane $x_n = 0$ in $\mathbb{R}^n$. Then stereographic projection from $x = (0, \ldots, 0, 1)$ sends the unit sphere to $\mathbb{R}^{n-1}_\infty \cup \{\infty\}$. This map extends to a Möbius transformation sending the unit ball to the upper half-space.

The metric on the sphere induced from $\mathbb{R}^n$ goes over to the metric $|dy|/\sigma(y)$ on $\mathbb{R}^{n-1}_\infty$, where

$$\sigma(y) = \frac{1 + y_1^2 + \ldots + y_{n-1}^2}{2}.$$

Now let t_i be the vector field $\partial/\partial x_i | S^{n-1}_\infty$, projected so it is tangent to the sphere. Then in the upper half-space model, we claim

$$t_1 = (\sigma - y_1^2, y_1 y_2, \ldots, y_1 y_{n-1})$$

(with respect to the basis $(\partial/\partial y_1, \ldots, \partial/\partial y_{n-1})$). The expression for t_i is similar, permuting coordinates.

To check this formula, note that t_1 represents the boundary values of an infinitesimal hyperbolic isometry, translating along the hyperbolic geodesic through $x = 0$ in the $\partial/\partial x_1$-direction. Thus t_1 vanishes at the intersections of the x_1-axis with the sphere, which correspond to $y = (\pm 1, 0, \ldots, 0)$. Also, t_1 points in the $\partial/\partial x_1$ direction with spherical length 1 at $x = (0, 0, \ldots, -1)$; equivalently, $t_1 = (1/2)\partial/\partial y_1$ when $y = 0$. These properties are all shared by the formula on the right and uniquely determine t_1.

The formula above generalizes the complex expression

$$t_1 = \frac{1}{2}(1 - z^2)\frac{\partial}{\partial z},$$

when $n = 2$ and $z = y_1 + iy_2$.

With these preliminaries in place, we can now explicitly calculate $\mathrm{av}(\omega)$ for some simple tensors.

Lemma B.5 *In the upper half-space model,*

$$\mathrm{av}\left(\frac{\partial}{\partial y_1}\right) = \frac{n}{2(n-1)}\frac{\partial}{\partial y_1}.$$

Proof. Let $v = \partial/\partial y_1$ on $\mathbb{R}^{n-1}_\infty$, and let $w = \partial/\partial y_1$ on $\mathbb{H}^n$. By symmetry considerations, $\mathrm{av}(v) = \alpha w$, so we need only compute

$$\alpha = \frac{\langle \mathrm{av}(v), w \rangle}{\langle w, w \rangle}$$

at any point in hyperbolic space. This expression is conformally invariant, so we can use the Euclidean inner product on the ball model and compute it at the origin $x = 0$. In the ball model, $w(0) = (1/2)\partial/\partial x_1$; thus $\langle w, w \rangle = 1/4$ and

$$\alpha = 2\left\langle \mathrm{av}(v)(0), \frac{\partial}{\partial x_1} \right\rangle = \frac{2}{\mathrm{vol}(S^{n-1}_\infty)} \int_{S^{n-1}_\infty} \left\langle v, \frac{\partial}{\partial x_1} \right\rangle dV.$$

Recall that t_1 is the vector field $\partial/\partial x_1$, projected so it is tangent to the sphere; thus α is twice the average of $\langle v, t_1 \rangle$ with respect to the spherical metric. Taking into account the conformal factor σ between the spherical metric and the Euclidean metric, we have $\langle v, t_1 \rangle = (\sigma - y_1^2)\sigma^{-2}$. Thus

$$\alpha = 2\left(\int_{\mathbb{R}^{n-1}_\infty} (\sigma - y_1^2) \frac{dy}{\sigma^{n+1}} \right) \Big/ \left(\int_{\mathbb{R}^{n-1}_\infty} \frac{dy}{\sigma^{n-1}} \right).$$

Here $dy = dy_1 \dots dy_n$. Integrating by parts allows one to express the numerator as a rational multiple of the denominator, and we find $\alpha = 2(n-1)/n$.

∎

A *Möbius vector field* is an infinitesimal Möbius transformation. On the circle, the Möbius vector fields are $sl_2(\mathbb{R})$; on spheres of higher dimension they are the same as the conformal vector fields.

Since $\mathrm{ex}(\cdot)$ is linear and the parabolic vector fields span the Möbius vector fields, we have:

Corollary B.6 *For any Möbius vector field v on S^{n-1}_∞, $\mathrm{ex}(v)$ is an isometric vector field on $\mathbb{H}^n$ and $v \cup \mathrm{ex}(v)$ is continuous.*

To analyze strain tensors, we need a similar calculation for the average of a symmetric tensor.

Lemma B.7 *In the upper half-space model,*

$$\text{av}\left(\frac{\partial}{\partial y_1} \otimes dy_1\right) = \left(\frac{n-2}{n}\right)\frac{\partial}{\partial y_1} \otimes dy_1 + \frac{I}{n(n-1)}.$$

Here I is the identity matrix.

Proof. Let $\mu = (\partial/\partial y_1) \otimes dy_1$ on $\mathbb{R}_\infty^{n-1}$; then μ is invariant under translations and real homotheties, so

$$\text{av}(\mu) = \sum \alpha_{ij} \frac{\partial}{\partial y_i} \otimes dy_j,$$

and to find the constants it suffices to calculate $\text{av}(\mu)$ at $x = 0$. By reasoning similar to that of the preceding proof, we find

$$\alpha_{ij} = \frac{1}{\text{vol}(S_\infty^{n-1})} \int_{S_\infty^{n-1}} \langle \mu t_i, t_j \rangle \, dV.$$

A straightforward integration by parts gives

$$\alpha_{11} = \left(\int_{\mathbb{R}_\infty^{n-1}} (\sigma - y_1^2)^2 \frac{dy}{\sigma^{n+1}}\right) \Big/ \left(\int_{\mathbb{R}_\infty^{n-1}} \frac{dy}{\sigma^{n-1}}\right) = \frac{n-2}{n} + \frac{1}{n(n-1)}.$$

For $i > 1$, we have

$$\alpha_{ii} = \left(\int_{\mathbb{R}_\infty^{n-1}} y_1^2 y_i^2 \frac{dy}{\sigma^{n+1}}\right) \Big/ \left(\int_{\mathbb{R}_\infty^{n-1}} \frac{dy}{\sigma^{n-1}}\right) = \frac{1}{n(n-1)};$$

and for $i \neq j$ the integrand is odd in one variable so $\alpha_{ij} = 0$.

$\blacksquare$

Corollary B.8 *If $\mu = \sum \alpha_{ij}(\partial/\partial y_i) \otimes dy_j$ is a translation-invariant strain field on $\mathbb{R}_\infty^{n-1}$, then $\mu \cup \text{ex}(\mu)$ is continuous on $S_\infty^{n-1} \cup \mathbb{H}^n$.*

Proof. Since α_{ij} is symmetric, μ is a linear combination of tensors equivalent to $(\partial/\partial y_1) \otimes dy_1$ up to rotations of $\mathbb{R}_\infty^{n-1}$. By the preceding lemma, $\text{ex}(\mu)$ and μ agree on S_∞^{n-1} apart from a multiple of I. But this multiple vanishes because $\text{tr}(\alpha_{ij}) = 0$.

$\blacksquare$

Proof of Theorem B.4 (Continuous extension). Let v be a continuous vector field, x a point in the sphere and p a point in hyperbolic space. Applying a rotation if necessary, we can assume $x \in \mathbb{R}^{n-1}_\infty$. As $p \to x$, the maximum visual length of v on any compact set disjoint from x tends to zero. Thus if $v(x) = 0$, then $\mathrm{ex}(v)(p) \to 0$, and $v \cup \mathrm{ex}(v)$ is continuous at x. But we can always find a translation-invariant vector field v_0 such that $(v - v_0)(x) = 0$. By Corollary B.6, $v_0 \cup \mathrm{ex}(v_0)$ is continuous, so $v \cup \mathrm{ex}(v)$ is also continuous at x for any continuous vector field v.

For a continuous strain field, the visual size of μ is independent of the choice of viewpoint p (since a strain tensor has degree zero). However, when p is near x, most geodesics through p land near x, so the limiting behavior of $\mathrm{ex}(\mu)$ still depends only on the germ of μ at x. Thus if $\mu(x) = 0$, then $\mathrm{ex}(\mu)(p) \to 0$ as $p \to x$. As before, we can correct μ by a translation-invariant strain field μ_0 so that $(\mu - \mu_0)(x) = 0$, and then continuity of $\mu \cup \mathrm{ex}(\mu)$ at x follows from the continuity of $\mu_0 \cup \mathrm{ex}(\mu_0)$ provided by Corollary B.8.

■

Even when v is smooth, we need to check the regularity of $v \cup \mathrm{ex}(v)$ further to assert that it is uniquely integrable on the closed ball. The following result is sufficient for our applications (cf. [Th1, Prop. 11.1.1]):

Proposition B.9 (Lipschitz extension) *If v is Lipschitz, then $v \cup \mathrm{ex}(v)$ is Lipschitz on $S^{n-1}_\infty \cup \mathbb{H}^n$.*

Proof. In the ball model for hyperbolic space, S^{n-1}_∞ is the unit sphere in $\mathbb{R}^n$, so we may regard v as a map from the sphere to $\mathbb{R}^n$. The Lipschitz condition means $|v(x) - v(x')| \leq M|x - x'|$ for some M, where $|\cdot|$ denotes the Euclidean norm on $\mathbb{R}^n$.

To show $v \cup \mathrm{ex}(v)$ is Lipschitz, pick $p \in \mathbb{H}^n$, $p \neq 0$; it suffices to show $|\mathrm{ex}(v)(p) - v(x)| \leq M'|p - x|$ where x is the radial projection of p to the sphere and M' is independent of p. Let v_0 be the unique parabolic vector field on $S^{n-1}_\infty \cup \mathbb{H}^n$ vanishing at $-x$ and with $v_0(x) = v(x)$. Since $|v(x)|$ is bounded, the Lipschitz constant of v_0 is bounded; moreover, v_0 is equal to the extension of its boundary values. Thus it suffices to show $|\mathrm{ex}(v - v_0)(p)| \leq M'|x - p|$. But $(v - v_0)$

is Lipschitz and vanishes at x, so $|(v-v_0)(x')| = O(|x-x'|)$. Thus the visual length of $(v - v_0)$ as seen from p is bounded, and therefore the hyperbolic length of $\mathrm{ex}(v-v_0)(p)$ is also bounded. The conformal factor between the Euclidean and hyperbolic metrics is approximately $|p - x|$ where p is near the sphere, so $|\mathrm{ex}(v - v_0)(p)| = O(|p - x|)$ as required.

∎

Strain energy. An invariant inner product between smooth, compactly supported vector fields on $\mathbb{H}^n$ is given by

$$\langle V, W \rangle = \int_{\mathbb{H}^n} \langle V(x), W(x) \rangle \, \rho^{n+2}(x) \, dx.$$

Here $\rho^n(x) \, dx$ is the hyperbolic volume form, and $\langle V(x), W(x) \rangle$ is the Euclidean inner product (which becomes hyperbolic when scaled by ρ^2).

Similarly, we define an inner product on strain fields by

$$\langle \mu, \nu \rangle \;=\; \int_{\mathbb{H}^n} \mathrm{tr}(\mu(x)\nu(x)) \, \rho^n(x) \, dx.$$

A function f on $\mathbb{H}^n$ is harmonic if it formally minimizes the Dirichlet energy $E(f) = \langle \nabla f, \nabla f \rangle$. Similarly, a vector field V is *strain-harmonic* if it formally minimizes the *strain energy*

$$E(V) = \langle SV, SV \rangle.$$

More precisely, we require that $\langle SV, SW \rangle = 0$ for every compactly supported smooth vector field W. Then

$$E(V + W) \;=\; E(V) + E(W) + 2\langle SV, SW \rangle \geq E(V)$$

for every compactly supported modification of V. (This calculation is only formal, since $E(V)$ may be infinite.)

Integrating by parts, we have

$$\langle SV, SW \rangle \;=\; \langle S_h^* SV, W \rangle \;=\; -\langle \rho^{-n-2} S^* \rho^n SV, W \rangle,$$

where

$$S^* \left(\sum \mu_{ij} \frac{\partial}{\partial x_i} \otimes dx_j \right) \;=\; \sum \frac{\partial \mu_{ij}}{\partial x_i} \frac{\partial}{\partial x_j}.$$

Thus V is strain-harmonic if and only if

$$S_h^* S V = 0.$$

When $n = 2$, this condition is equivalent to $\partial \rho^2 \overline{\partial} V = 0$.

The notation S_h^* indicates the *hyperbolic* adjoint to S (as opposed to the Euclidean adjoint S^*). Since S is natural, the map

$$\mu \mapsto S_h^*(\mu) \;=\; \rho^{-n-2} S^* \rho^n \mu$$

(from smooth strain fields on $\mathbb{H}^n$ to smooth vector fields on $\mathbb{H}^n$) is also natural with respect to hyperbolic isometries.

We have seen that the visual extension of a function on the sphere is harmonic; similarly, the visual extension of a vector field is strain-harmonic.

Theorem B.10 (Quasiconformal extension) *Let v be a quasi-conformal vector field on S_∞^{n-1}, $n \geq 3$, and let $V = \mathrm{ex}(v)$. Then V is a smooth, volume-preserving quasi-isometric vector field, continuously extending v and formally minimizing strain energy. Moreover*

$$SV = \mathrm{ex}(Sv),$$

so $\|SV\|_\infty \leq (n/(n-2))\|Sv\|_\infty$.

Proof. We have already seen that V is a smooth vector field continuously extending v. The operator $v \mapsto \nabla \cdot \mathrm{ex}(v)$ is a natural map from vector fields to functions, so by Theorem B.1 it is zero; thus V is volume preserving. Similarly, $S(\mathrm{ex}(v))$ and $\mathrm{ex}(Sv)$ can only differ by a real multiple; but for $v = (y_1, -y_2, 0, \ldots, 0)$ on $\mathbb{R}_\infty^{n-1}$, both are continuous extension of Sv, so they are equal. Clearly

$$\|SV(0)\| \;=\; \frac{n}{n-2}\|\,\mathrm{av}(Sv)(0)\| \;\leq\; \frac{n}{n-2}\|Sv\|_\infty;$$

since $\mathrm{av}(\cdot)$ is natural and the norm of a strain tensor is conformally invariant, the same bound on SV holds throughout hyperbolic space.

Thus V is quasiconformal; since it is also volume-preserving, it is actually a quasi-isometric vector field.

The operator $\mu \mapsto S_h^* \mathrm{ex}(\mu)$ is a natural map from smooth strain fields on S_∞^{n-1} to smooth vector fields on $\mathbb{H}^n$, so it is zero by Theorem

B.1. By continuity, the same is true when μ is a distributional strain field. But then

$$S_h^* S(\text{ex}(v)) = S_h^*(\text{ex}(Sv)) = 0,$$

so $\text{ex}(v)$ is strain-harmonic.

∎

Remark. Since $\nabla \cdot V = 0$, the conformal strain accounts for the total distortion of the metric under $V = \text{ex}(v)$, so $\text{ex}(v)$ is also an infinitesimal harmonic map.

Now let $S^n = \mathbb{R}^n \cup \{\infty\}$, and let $\overline{\text{ex}(v)}$ denote the pushforward of $\text{ex}(v)$ by reflection through S_∞^{n-1}.

Corollary B.11 *If v is a k-quasiconformal vector field on S_∞^{n-1}, then $V = v \cup \text{ex}(v) \cup \overline{\text{ex}(v)}$ is a k'-quasiconformal vector field on S^n, where $k' = nk/(n-2)$.*

Proof. The only possible difficulty is that the distribution SV might be singular along S_∞^{n-1}. To rule this out, as in the proof of Theorem A.1 we express v as the uniform limit of smooth vectors fields $v^i = \alpha_i * v$, for a sequence of smooth probability measures α_i on $SO(n)$ converging to the δ-mass at the identity. Since v^i is Lipschitz, $V^i = v^i \cup \text{ex}(v^i) \cup \overline{\text{ex}(v^i)}$ is a Lipschitz vector field on S^n (Proposition B.9). Thus V^i has derivatives locally in L^∞, and

$$\|SV^i\|_\infty \leq \frac{n}{n-2}\|Sv^i\|_\infty \leq \frac{n}{n-2}\|Sv\|_\infty \leq \frac{nk}{n-2}.$$

Passing to the limit as $i \to \infty$, we conclude V is k'-quasiconformal.

∎

Fundamental solutions. Another approach to the properties of $\text{ex}(v)$ is to observe that if $v = \delta(x-N)\partial/\partial x_1$ is a distributional vector field on the sphere, concentrated at $N = (0,0,\dots,1)$, then $\text{ex}(v)$ is a positive multiple of $y_n^{n+1}\partial/\partial y_1$. This "fundamental solution" is evidently volume-preserving (it is a shear), and one can also check that it is strain-energy minimizing. Since the linear span of the orbit of v under $\text{Isom}(\mathbb{H}^n)$ is dense in the space of distributional vector fields, these properties are inherited by $\text{ex}(v)$ for all vector fields.

B.2   Representation theory

In this section we prove Theorem B.1. The proof will use some basic facts from representation theory, which we now recall.

Definitions. A *representation* of a Lie group G is a continuous linear action of G on a complete topological vector space V. We refer to V as a G-*module*; the G-action will be written $(g, v) \mapsto g \cdot v$. A G-module is *irreducible* if $\dim V > 0$ and V contains no closed invariant subspace other than $\{0\}$ and V. Given a pair of G-modules, $\text{Hom}_G(V_0, V_1)$ is the space of continuous linear maps $T : V_0 \to V_1$ which respect the action of G. For finite-dimensional irreducible representations over $\mathbb{C}$, *Schur's Lemma* states that $\text{Hom}_G(V_0, V_1)$ is one-dimensional if V_0 and V_1 are isomorphic G-modules, and zero-dimensional otherwise.

Let $X = G/K$ be a homogeneous space, where K is a closed subgroup of G. Given a finite-dimensional K-module V, we obtain a vector bundle $E(V) \to X$ by setting

$$E(V) \;=\; G \times_K V \;=\; (G \times V)/\{(gk, v) \sim (g, k \cdot v) \text{ for all } k \in K\}.$$

The action of G on G/K is covered by a natural action on $E(V)$, so we obtain an action of G on the space

$$\text{Ind}_K^G(V) \;=\; C^\infty(X, E(V))$$

of smooth sections of $E(V)$. This G-module is *induced* from K to G.

An equivalent description of $\text{Ind}_K^G(V)$ is the following: it is the space of smooth maps $f : G \to V$ such that $f(g) = k \cdot f(gk)$ for all $k \in K$. The action of G is given by $(g_0 \cdot f)(g) = f(g_0^{-1} g)$.

Theorem B.12 (Frobenius reciprocity) *Let V and W be finite-dimensional modules over K and G respectively. Then there is a natural isomorphism*

$$\text{Hom}_G(W, \text{Ind}_K^G(V)) \cong \text{Hom}_K(W, V).$$

Proof. Let α belong to $\text{Hom}_G(W, \text{Ind}_K^G(V))$. Thinking of $\alpha(w)$ as a smooth map from G to V, we obtain a linear map $\beta : W \to V$ by setting $\beta(w) = \alpha(w)(\text{id})$. It is easy to see that β respects the action of K, so it belongs to $\text{Hom}_K(W, V)$. The mapping $\beta \mapsto \alpha$ given by $\alpha(w)(g) = \beta(g \cdot w)$ inverts this correspondence, so we have an isomorphism.

∎

Proposition B.13 *If, in addition, G is compact, then any natural map $\alpha : C^\infty(X, E(V)) \to W$ is given by pairing with a smooth kernel. Thus α extends continuously to a map from $\mathcal{D}(X, E(V))$ to W, and we may construct an isomorphism*

$$\text{Hom}_G(\text{Ind}_K^G(V), W) \cong \text{Hom}_G(W, \text{Ind}_K^G(V)).$$

Proof. Since G is compact, there are G-invariant inner products on V and W, and there is a finite smooth G-invariant measure dx on $X = G/K$. Thus we have an invariant inner product

$$\langle \sigma_1, \sigma_2 \rangle = \int_X \langle \sigma_1(x), \sigma_2(x) \rangle \, dx$$

on the space of smooth sections of $E(V)$ over X, and we can identify $\text{Ind}_G^K(V)^*$ with the space of distributional sections $\mathcal{D}(X, E(V))$. By dualizing, any G-map $\alpha : \text{Ind}_K^G(V) \to W$ determines a map

$$\alpha^* : W \cong W^* \to \mathcal{D}(X, E(V)).$$

Now for any smooth measure $\mu = \mu(g)dg$ on G, and any distributional section ϕ of $E(V)$, the convolution $\mu * \phi$ is a smooth section of $E(V)$. If μ is concentrated sufficiently close to the identity, then convolution with μ sends a basis for W^* to another basis. But $\alpha^*(\mu * w) = \mu * \alpha^*(w)$, so α^* actually maps W to the space of smooth sections $\text{Ind}_K^G(V)$.

Dualizing once more, we find α^{**} is an extension of α to the space of distributional sections. In particular, $\alpha \mapsto \alpha^*$ is an isomorphism from $\text{Hom}_G(\text{Ind}_K^G(V), W)$ to $\text{Hom}_G(W, \text{Ind}_K^G(V))$.

∎

To apply these ideas to hyperbolic space, we need some facts about the orthogonal group. For $n, d \geq 0$, let $P_d(n)$ denote the space of real-valued polynomials on $\mathbb{R}^n$, homogeneous of degree d. Let $H_d(n) \subset P_d(n)$ be the *harmonic polynomials* (those annihilated by the Laplacian $\Delta = \partial^2/\partial x_1^2 + \ldots + \partial^2/\partial x_n^2$).

The space $H_d(n) \otimes \mathbb{C}$ is an irreducible $O(n)$-module whenever its dimension is positive. These irreducible representations are distinct as d varies. We have an invariant splitting

$$P_d(n) = H_d(n) \oplus |x|^2 P_{d-2}(n),$$

which implies

$$\dim H_d(n) = \dim P_d(n) - \dim P_{d-2}(n) = \binom{n+d-1}{d} - \binom{n+d-3}{d-2}$$

(see Table B.1). For proofs of these basic facts see, e.g. [Hel, p.16].

$\dim H_d(n)$	$d = 0$	$d = 1$	$d = 2$	$d = 3$	$d = 4$
$n = 0$	1	0	0	0	0
$n = 1$	1	1	0	0	0
$n = 2$	1	2	2	2	2
$n = 3$	1	3	5	7	9
$n = 4$	1	4	9	16	25

Table B.1. Spaces of harmonic polynomials

Next we examine how $H_d(n)$ decomposes as an $O(n-1)$-module under the standard inclusion $O(n-1) \subset O(n)$.

Proposition B.14 *For $n \geq 2$,*

$$H_d(n) \cong H_0(n-1) \oplus H_1(n-1) \oplus \cdots \oplus H_d(n-1)$$

as an $O(n-1)$ *module. Consequently, when both spaces have positive dimension,*

$$\dim \operatorname{Hom}_{O(n-1)}(H_d(n), H_e(n-1)) = \begin{cases} 1 & e \leq d, \\ 0 & otherwise. \end{cases}$$

Proof. Given $p \in H_d(n)$, write $p = \sum_{e=1}^d p_e(x) x_n^{d-e}$, where $p_e \in P_e(n-1)$ is independent of x_n. Define $T : H_d(n) \rightarrow \bigoplus_1^d H_e(n-1)$ by projecting $p_e(x)$ to the harmonic polynomials of degree e for each e. Clearly T respects the action of $O(n-1)$. Suppose $p \neq 0$, and consider the least e such that $p_e \neq 0$. Then the part of Δp of highest degree in x_n is $(\Delta p_e) x_n^{d-e}$. Since p is harmonic, we conclude $\Delta p_e = 0$, and thus $T(p) \neq 0$. This shows T is injective. Surjectivity follows by counting dimensions.

Since the irreducible representations $H_e(n-1)$ are distinct as e varies, and each occurs with multiplicity at most one in $H_d(n)$, the last part of the Theorem follows over $\mathbb{C}$ by Schur's Lemma. But it is also true over $\mathbb{R}$, since the map T is defined over $\mathbb{R}$.

∎

Proof of Theorem B.1 (Natural maps). Let $G = O(n)$ and $K = O(n-1)$, so $X = G/K = S_\infty^{n-1}$. Let $N = (0, 0, \dots, 1) \in \mathbb{R}^n$ be the "north pole" of the sphere; then K is the stabilizer of N, and so it acts on the tangent space to N. Thus $\mathbb{R}$, $\mathrm{T}_N(S_\infty^{n-1})$ and $\mathrm{sym}_0(\mathrm{T}_N(S_\infty^{n-1}))$ are K-modules; in fact, they are isomorphic to $H_d(n-1)$ for $d = 0, 1, 2$. Consequently, $V_d = \mathrm{Ind}_K^G(H_d(n-1))$ is isomorphic to $F(S_\infty^{n-1})$, $V(S_\infty^{n-1})$ or $S(S_\infty^{n-1})$, when $d = 0$, 1 or 2. Similarly, G is the stabilizer of the origin $x = 0$ in hyperbolic space, and the G-modules $\mathbb{R}$, $\mathrm{T}_0(\mathbb{H}^n)$ and $\mathrm{sym}_0(\mathrm{T}_0(\mathbb{H}^n))$ are isomorphic to $H_0(n)$, $H_1(n)$ and $H_2(n)$. Denote $F(\mathbb{H}^n)$, $V(\mathbb{H}^n)$ and $S(\mathbb{H}^n)$ by W_0, W_1 and W_2.

For $0 \leq d, e \leq 2$, let $\alpha : V_e \rightarrow W_d$ be a map which is natural with respect to $\mathrm{Isom}(\mathbb{H}^n)$. Then using the identifications above, $\beta(\omega) = \alpha(\omega)(0)$ is a G-map from V_e to $H_d(n)$. Conversely, any G-map $\beta : V_e \rightarrow H_d(n)$ prolongs to a natural map $\alpha : V_e \rightarrow W_d$ by setting $\alpha(\omega)(g0) = g_* \beta(\omega)(0)$. Indeed, by Proposition B.13, β is given by pairing with a smooth kernel, so the same is true of α. This shows

$\alpha(\omega)$ is smooth on $\mathbb{H}^n$ and also provides an extension of α to the space of distributional sections.

Applying Frobenius reciprocity and Proposition B.13, we find the space of natural maps $\alpha : V_e \to W_d$ has the same dimension as $\mathrm{Hom}_K(H_d(n), H_e(n-1))$. By the preceding Proposition, this dimension is 1 for $d \geq e$ and zero otherwise (as long as both $H_d(n)$ and $H_e(n-1)$ have positive dimension). As d and e vary we obtain the nine cases stated in Theorem B.1.

$\blacksquare$

Remark. A vector field $v(z)\partial/\partial z$ on $S_\infty^2 = \widehat{\mathbb{C}}$ can be rotated by a complex scalar α. Using this freedom, we obtain additional mappings $v \mapsto \mathrm{av}(\alpha v)$ which are natural with respect to $\mathrm{Isom}^+(\mathbb{H}^3)$. However, these maps commute with orientation-reversing isometries only when α is real, so Theorem B.1 remains valid.

B.3   The visual distortion

A vector field v on the sphere is quasiconformal if and only if $\|Sv\|_\infty < \infty$. In this section we introduce a more geometric measure of quasiconformality, the *visual distortion* $Mv(p)$. This function measures the visual distance from v to the conformal vector fields, as seen from $p \in \mathbb{H}^n$. The visual distortion plays a pivotal role in establishing "effective rigidity" for hyperbolic manifolds and conformal dynamical systems (see §2.4 and §9.3).

It is easy to see v is quasiconformal if and only if Mv is bounded. The visual distortion has the virtue of measuring the behavior of v at different scales; in particular, Mv can detect that v is uniformly close to a vector field whose distortion is much less than $\|Sv\|_\infty$. We develop several estimates and a maximum principle based on these ideas.

Definitions. Let v be a vector field on S_∞^{n-1}, and let p be a point in hyperbolic space $\mathbb{H}^n$. We let ρ_p denote the visual metric on S_∞^{n-1} as seen from p. In the ball model, the visual length $\rho_0(v(x))$ agrees with the Euclidean length $|v(x)|$ induced from $\mathbb{R}^n$. The visual metric is defined at other points $p = g(0)$ by naturality $(\rho_p(g_*v) = \rho_0(v))$.

The *visual maximum* of v as seen from p is given by

$$\|v\|_\infty(p) = \sup_{x \in S_\infty^{n-1}} \rho_p(v(x)).$$

Thus the Euclidean maximum $\|v\|_\infty$ is the same as $\|v\|_\infty(0)$. The *visual distortion* is defined by

$$Mv(p) = \inf_{Sw=0} \|v - w\|_\infty(p).$$

The infimum is over all conformal vector fields w on S_∞^{n-1}. The visual distortion is the apparent size of v as seen from p, after correcting by a conformal vector field to make the size as small as possible.

Theorem B.15 *For any continuous vector field v on S_∞^{n-1},*

$$\frac{1}{C_n} \|Sv\|_\infty \leq \sup_{\mathbb{H}^n} Mv(p) \leq C_n \|Sv\|_\infty.$$

Thus v is quasiconformal if and only if the visual distortion $Mv(p)$ is uniformly bounded on hyperbolic space. We also have the pointwise bound

$$\|SV(p)\| \leq C_n Mv(p),$$

where $V = \mathrm{ex}(v)$.

Proof. We begin by bounding $\|SV(0)\|$. Since SV is unchanged by adding a conformal vector field w to v, we may as well assume $\|v\|_\infty \leq Mv(0)$. The extension operator is continuous from the space of distributional vector fields on the sphere to the space of C^∞ vector fields on hyperbolic space (Theorem B.1), so a bound on $\|v\|_\infty$ gives a bound on V and its derivatives at the origin in hyperbolic space. Thus $\|SV(0)\| \leq C_n Mv(0)$, and by naturality the same bound holds at every point $p \in \mathbb{H}^n$.

Consequently $\|SV\|_\infty \leq C_n \|Mv\|_\infty$. Now if v is smooth, then SV is a continuous extension of Sv, and thus $\|Sv\|_\infty \leq \|SV\|_\infty$. As we have seen, a continuous vector field can be approximated by a smooth one by convolving with a smooth measure on the rotation group, so $\|Sv\|_\infty \leq C_n \|Mv\|_\infty$ for general v.

The reverse inequality follows from Theorem A.10, which states that a quasiconformal vector field on the sphere can be corrected by a conformal vector field so $\|v\|_\infty$ is bounded in terms of $\|Sv\|_\infty$. This shows $Mv(0) \leq C_n\|Sv\|_\infty$; by naturality, the same is true at every point in hyperbolic space, so $\|Mv\|_\infty \leq C_n\|Sv\|_\infty$. ∎

Remark. A bound on the visual distortion Mv provides much more control over v than the corresponding bound on SV, where $V = \mathrm{ex}(v)$. For example, we have just shown $SV(p) = O(Mv(p))$. On the other hand, there exist vector fields v_i with $Mv_i(p) \to \infty$, but with $SV_i \to 0$ uniformly on compact subsets of $\mathbb{H}^n$.

Let $S(p, r)$ denote the hyperbolic sphere of radius r about $p \in \mathbb{H}^n$. The next Proposition says *any* vector field that has bounded visual distortion on a large hyperbolic sphere is well-approximated by a k-quasiconformal vector field, as seen from the center of the sphere.

Theorem B.16 *If $Mv(q) \leq 1$ for all $q \in S(p, r)$, then there is a k-quasiconformal vector field v_0 on S_∞^{n-1} such that*

$$\|v - v_0\|_\infty(p) \leq C_n e^{-r}.$$

Here k depends only on the dimension n.

Proof. By naturality we can assume $p = 0$ in the ball model. Let us translate the hypothesis of the Theorem into bounds in the Euclidean metric.

The Euclidean distance from $S(0, r)$ to S_∞^{n-1} is comparable to $\epsilon = e^{-r}$ when r is large. Each point $x \in S_\infty^{n-1}$ projects radially to a point $q \in S(0, r)$, and the conformal factor between ρ_q and the Euclidean metric ρ_0 is approximately $1/\epsilon$ on an ϵ-ball around x. Since $Mv(q) \leq 1$, there is a conformal vector field w such that $|v - w| = O(\epsilon)$ on $B(x, \epsilon)$. By compactness the unit sphere is covered by a finite collection of ϵ-balls B_i, each equipped with a conformal vector field w_i such that $|v - w_i| = O(\epsilon)$ on B_i. We can also assume any point on the sphere belongs to at most N balls, where N depends only on n.

Choose a smooth partition of unity ρ_i, with ρ_i supported on B_i, $\sum \rho_i = 1$ and $|\nabla \rho_i(x)| = O(1/\operatorname{diam} B_i) = O(1/\epsilon)$. Let $v_0 = \sum \rho_i w_i$. Clearly $\|v(x) - v_0(x)\| = O(\epsilon)$. It remains to estimate Sv_0. For

$x \in B_i$, we can replace v_0 by $v_0 - w_i$ without changing its strain. Then

$$\|Sv_0(x)\| = O\left(\sum_{B_j \cap B_i \neq \emptyset} |\nabla \rho_j(x)| \cdot |w_j(x) - w_i(x)| \right)$$
$$= O(N \cdot \epsilon^{-1} \cdot \epsilon) = O(1).$$

Thus v_0 is k-quasiconformal for a universal k, completing the proof.

∎

Corollary B.17 *The visual distortion satisfies the following maximum principle: for any ball B in hyperbolic space,*

$$\sup_B Mv(p) \leq C_n \sup_{\partial B} Mv(p).$$

Proof. By homogeneity we can assume $\sup_{\partial B} Mv(p) = 1$. Then it suffices to show $Mv(p) = O(1)$ for all $p \in B$.

Let c and $r > 0$ be the center and radius of B. The preceding Theorem gives $v = v_0 + v_1$ where v_0 is k-quasiconformal and $Mv_1(c) \leq e^{-r}$. Since $\rho_p/\rho_c = O(e^r)$ for any $p \in B$, we have $Mv_1(p)/Mv_1(c) = O(e^r)$ and thus $Mv_1(p) = O(1)$. Since v_0 is k-quasiconformal, $Mv_0(p) = O(1)$ also. Thus $Mv(p) = O(1)$; more precisely, $Mv(p)$ is bounded by a constant depending only on n.

∎

Corollary B.18 *Suppose v_i is a sequence of continuous vector fields on S_∞^{n-1}, such that $Mv_i(q) \leq 1$ on the sphere of radius $r_i \to \infty$ about $p \in \mathbb{H}^n$. Then after passing to a subsequence and correcting by conformal vector fields, v_i converges uniformly to a quasiconformal vector field v.*

Proof. By Theorem B.16 there are k-quasiconformal vector fields u_i such that $\|v_i - u_i\|_\infty(p) \leq \epsilon_i \to 0$ as $i \to \infty$. By compactness of k-quasiconformal vector fields (Corollary A.11), there are conformal vector fields w_i such that $u_i + w_i$ converges uniformly to a k-quasiconformal vector field v_∞. Then the same is true of $v_i + w_i$.

∎

Remarks. The vector fields v_i in the preceding Corollary can be wild on a small scale (for example, they need not be quasiconformal). However, this scale tends to zero as $i \to \infty$.

The visual L^λ norm. If one has a bound on both $\|Sv\|_\infty$ and the size of the support of Sv, then one can improve the estimate for the visual distortion. This improvement is conveniently phrased in terms of the L^λ norm, where $(n-1) < \lambda < \infty$.

Let v be a continuous vector field on S^{n-1}_∞ with strain Sv in L^λ. The L^λ norm of v depends on a choice of measure. For each $p \in \mathbb{H}^n$, let $dV_p(x)$ denote the measure on the sphere corresponding to the visual metric ρ_p, scaled to have total mass one. In the ball model dV_0 is a multiple of the $(n-1)$-dimensional Lebesgue measure induced from $\mathbb{R}^n$.

The *visual L^λ norm* of the strain Sv is defined by

$$\|Sv\|_\lambda(p) = \left(\int_{S^{n-1}_\infty} \|Sv(x)\|^\lambda \, dV_p(x) \right)^{1/\lambda}.$$

By Theorem A.13 we have:

Theorem B.19 *For any $p \in \mathbb{H}^n$, and $(n-1) < \lambda \leq \infty$, we have*

$$Mv(p) \leq C_{\lambda,n} \|Sv\|_\lambda(p).$$

For example, setting $\lambda = n$ one obtains:

Corollary B.20 *If v is a quasiconformal vector field, and $Sv = 0$ outside a measurable set $E \subset S^{n-1}_\infty$, then*

$$Mv(p) \leq C_n \|Sv\|_\infty \cdot (\text{visual measure of } E \text{ as seen from } p)^{1/n}.$$

The case of $\widehat{\mathbb{C}}$: cross-ratios and quadratic differentials. It is clear that Mv depends only on Sv, because if $Sv_1 = Sv_2$ then $v_1 - v_2$ is a conformal vector field. Thus it is natural to try to express Mv directly in terms of $\mu = Sv$, at least up to a bounded factor. In dimension two this can be done fairly explicitly.

First, call a quadruple of points $\{a, b, c, d\}$ on the Riemann sphere *well-separated* (relative to $p \in \mathbb{H}^3$) if the visual distance between any two distinct points is at least $1/10$. Let w be the unique conformal vector field such that $v(x) = w(x)$ when $x = a$, b or c. Then the visual length of $(v - w)(d)$ measures the distortion of the cross-ratio $[a : b : c : d]$ under v. It is not hard to show that $Mv(p)$ is comparable to the supremum of the cross-ratio distortion over all well-separated quadruples. But the cross-ratio distortion is itself comparable to

$$\left| \int_{\widehat{\mathbb{C}}} \mu\phi \right| \bigg/ \int_{\widehat{\mathbb{C}}} |\phi|,$$

where ϕ is the quadratic differential $dz^2/((z-a)(z-b)(z-c)(z-d))$ (and $\mu = \bar{\partial}v$). One way to check this is to integrate by parts, and use the fact that $\bar{\partial}\phi$ is a distribution concentrated at $\{a, b, c, d\}$. More conceptually, μ represents a tangent vector to the Teichmüller space of the four-times punctured sphere $\widehat{\mathbb{C}} - \{a, b, c, d\}$, and the ratio above is exactly the Teichmüller length of this vector (cf. [Gard]). For well-separated quadruples, the cross-ratio distortion is comparable to this Teichmüller length.

The case of S^1. It is natural to define the visual distortion of a vector field on the circle by simply replacing $Sw = 0$ with the condition that w is a Möbius vector field (an element of $\text{sl}_2(\mathbb{R})$). Then the vector fields with $Mv(p)$ bounded on $\mathbb{H}^2$ are exactly those which satisfy the Zygmund condition; they are also characterized by $\|SV\|_\infty < \infty$, where $V = \text{ex}(v)$. These vector fields generate quasisymmetric flows and form the natural class of quasiconformal vector fields in dimension one (cf. [AK]).

B.4   Extending quasiconformal isotopies

We now apply the infinitesimal deformation theory described above to extend a quasiconformal isotopy of the sphere to a quasi-isometric isotopy of hyperbolic space. In our primary case of interest, extension from S^2_∞ to $\mathbb{H}^3$, any quasiconformal map can be included in an isotopy by solving the Beltrami equation. As a Corollary, we obtain a quasi-isometry between the quotient manifolds of any pair of quasiconformally conjugate Kleinian groups.

First assume $n \geq 3$.

Theorem B.21 (Visual isotopy extension) *Let $\phi : S^{n-1}_\infty \times [0, 1] \to S^{n-1}_\infty$ be an isotopy of the sphere, obtained by integrating a continuous k-quasiconformal vector field $v : [0, 1] \times S^{n-1}_\infty \to \text{T}S^{n-1}_\infty$; then:*

1. *There is a unique isotopy*

$$\Phi : \mathbb{H}^n \times [0,1] \to \mathbb{H}^n$$

   obtained by integrating the visual extension $V_t = \mathrm{ex}(v_t)$.

2. *For each* t, Φ_t *is a volume-preserving* L_t-*quasi-isometry of hyperbolic space, where* $L_t = \exp(nkt/(n-2))$.

3. *The map* $\phi \cup \Phi$ *is an isotopy of the closed ball* $S_\infty^{n-1} \cup \mathbb{H}^n$.

4. *The pointwise quasi-isometry constant* $L(\Phi_t, x)$ *satisfies*

$$\log L(\Phi_t, x) \; \leq \; \int_0^t \|SV_s(\Phi_s(x))\| \, ds.$$

5. *The extension is natural in the sense that for any family* g_t *of hyperbolic isometries such that*

$$g_t \circ \phi_t = \phi_t \circ g_0$$

   on the sphere, we also have $g_t \circ \Phi_t = \Phi_t \circ g_0$ *on hyperbolic space.*

Proof. Let $W_t = v_t \cup \mathrm{ex}(v_t) \cup \overline{\mathrm{ex}(v_t)}$, where $\overline{\mathrm{ex}(v_t)}$ is $\mathrm{ex}(v_t)$ reflected through S_∞^{n-1}. By Corollary B.11, W_t is a continuous, k'-quasiconformal vector field on S^n, where $k' = nk/(n-2)$. By Proposition B.3, a uniformly small change in v_t makes a small change in $\mathrm{ex}(v_t)$; thus W is continuous on $S^n \times [0,1]$. By Theorem A.1, there is a unique quasiconformal flow $\Psi : S^n \times [0,1] \to S^n$ integrating W, and the dilatation $K(\Psi_t)$ is bounded by $\exp(nk't)$.

Since W_t is tangential to S_∞^{n-1}, Ψ_t maps hyperbolic space into itself, so by restriction we obtain a flow $\Phi : \mathbb{H}^n \times [0,1] \to \mathbb{H}^n$ integrating V. Theorem B.10 implies Φ_t preserves hyperbolic volume, so its quasi-isometric distortion is bounded by the nth root of its quasiconformal distortion, $K(\Psi_t)^{1/n} \leq \exp(k't) = \exp(nkt/(n-2))$. The pointwise bound follows by considering the family of metrics $rho_t = \Phi_t^* rho_0$, where rho_0 is the hyperbolic metric, and applying equation (A.4).

Finally we check naturality. Let g_t be a family of automorphisms of the closed ball $S_\infty^{n-1} \cup \mathbb{H}^n$ such that $g_t \circ \phi_t = \phi_t \circ g_0$. Then

$$\kappa_t = \frac{d}{ds} g_s \circ g_t^{-1} \Big|_{s=t}$$

is vector field which is conformal on the sphere and isometric on hyperbolic space. Substituting $g_s = \phi_s \circ g_0 \circ \phi_s^{-1}$, we find

$$\kappa_t | S_\infty^{n-1} = v_t - (g_t)_*(v_t).$$

(This equation expresses κ_t as the coboundary of v_t.) Applying $\mathrm{ex}(\cdot)$ to this equation, we have

$$\kappa_t | \mathbb{H}^n = V_t - (g_t)_*(V_t).$$

But the right-hand side of this equation is the derivative of $\Phi_t \circ g_0 \circ \Phi_t^{-1}$ at time t. Thus $\Phi_t \circ g_0 \circ \Phi_t^{-1} = g_t$.

∎

Remarks. Since the visual extension of a vector field is smooth, one may wonder if the more delicate results on unique integrability are really needed in the proof above. In fact these results play two roles: first, to assure that a flow line for the extended vector field does not reach the edge of $\mathbb{H}^n$ in finite time; and second, to show that the flow on $\mathbb{H}^n$ extends the given flow on S_∞^{n-1}. Note that the non-uniquely integrable vector field in Figure A.1 of §A.5 is smooth in the upper and lower half-planes, and tangential to the real axis. Nevertheless it admits a flow line crossing from one half-plane to the other.

The Beltrami isotopy. Let $f : S_\infty^2 \to S_\infty^2$ be a quasiconformal map fixing 0, 1 and ∞. Then f can be included in a quasiconformal isotopy by solving the Beltrami equation. To make this precise, identify S_∞^2 with the Riemann sphere $\hat{\mathbb{C}}$, let $\mu = (f_{\bar{z}}/f_z)d\bar{z}/dz$ be the complex dilatation of f, let $\tau = \|\mu\|_\infty$ and let $\nu = (1/\tau)\mu$. Then there is a unique map $\psi : \Delta \times \hat{\mathbb{C}} \to \hat{\mathbb{C}}$, sending (λ, z) to $\psi_\lambda(z)$, such that:

1. ψ is continuous;

2. for fixed λ, $z \mapsto \psi_\lambda(z)$ is a quasiconformal map with complex dilatation $\lambda\nu$, fixing 0, 1 and ∞;

3. for fixed z, $\lambda \mapsto \psi_\lambda(z)$ is holomorphic;

4. $d\psi_\lambda(z)/d\lambda = v_\lambda(\psi_\lambda(z))$, where $v_\lambda(z)$ is a continuous, quasiconformal vector field with $\|\bar{\partial}v_\lambda\|_\infty = 1/(1 - |\lambda|^2)$; and

5. $\psi_\tau(z) = f(z)$.

For a proof, see [AB].

It is convenient to reparameterize the isotopy ψ by hyperbolic arc length along the interval $[0, \tau]$, scaled so the total length is one. After reparameterization, we obtain an isotopy ϕ_t with a constant bound on $\|\bar{\partial} v_t\|_\infty$ instead of a bound which depends on t. The complex dilatation of ϕ_t is $\tanh(tk)\nu$. In summary we have:

Theorem B.22 (Beltrami isotopy) *Let* $f : S^2_\infty \to S^2_\infty$ *be a K-quasiconformal map fixing 0, 1 and ∞, and let $K = e^{2k}$. Then there is a quasiconformal isotopy*

$$\phi : S^2_\infty \times [0, 1] \to S^2_\infty,$$

also fixing 0, 1 and ∞, obtained by integrating a continuous k-quasiconformal vector field v, such that $\phi_0 = \mathrm{id}$ and $\phi_1 = f$.

The isotopy is natural in the sense that for any Möbius transformations γ_0, γ_1 such that

$$\gamma_1 \circ f = f \circ \gamma_0,$$

there are interpolating Möbius transformations γ_t such that

$$\gamma_t \circ \phi_t = \phi_t \circ \gamma_0.$$

Applying Theorem B.21 to the Beltrami isotopy, we have:

Corollary B.23 *Let* $M_i = \mathbb{H}^3/\Gamma_i$, $i = 1, 2$ *be a pair of hyperbolic 3-manifolds, and let ϕ be a K-quasiconformal conjugacy between Γ_0 and Γ_1. Then ϕ extends to an equivariant L-quasi-isometry $\Phi : \mathbb{H}^3 \to \mathbb{H}^3$, where $L = K^{3/2}$. In particular, M_1 and M_2 are quasi-isometric manifolds.*

Remarks.

1. Applying the Corollary to the trivial group, we see any quasiconformal map on $\hat{\mathbb{C}}$ extends to a quasi-isometry of hyperbolic space.

2. The Beltrami isotopy shows a quasiconformal map on the Riemann sphere can be factored in a natural way into maps of small dilatation. (Such a factorization result is unknown in higher dimensions.)

3. It is worth noting that there are K-quasiconformal mappings f_n fixing 0, 1 and ∞, such that the complex dilatation μ_n of f_n converges weak* to zero, but f_n does not converge to the identity mapping. One such family is given by $f_n(x, y) = (x, g_n(y))$, where $g'_n(y) = K$ when $y \in (i/n, (i+1)/n)$ and i is even, and $g'_n(y) = 1/K$ when i is odd. The mapping f_n sends horizontal strips of height $1/n$ to strips of heights alternating K/n and $1/(nK)$. The complex dilatation alternates sign from strip to strip, so it tends weakly to zero; but $f_n(x, y) \rightarrow (x, Ay)$ where $A = (K + 1/K)/2$.

Because of this phenomenon, it is not possible to estimate the terminal mapping ϕ_1 of a Beltrami isotopy in terms of the size of the vector field $v_0 = d\phi_t/dt|_{t=0}$. That is, even if the spherical length of v_0 is small, the distance from ϕ_1 to the identity map can be large.

On the other hand, if $\|\mu\|_\infty = 1$ and $\psi_\lambda(z)$ is the unique normalized map with complex dilatation $\lambda\mu$, then we do have

$$d(z, \psi_\lambda(z)) \leq C(|\lambda| \cdot \|v(z)\| + |\lambda|^2)$$

for a universal constant C. Here d and $\|v\|$ are measured in the spherical metric. This estimate follows from the Schwarz lemma and the fact that $\lambda \mapsto \psi_\lambda(z)$ is a holomorphic map from the unit disk into $\hat{\mathbb{C}} - \{0, 1, \infty\}$. So when $\|v\|_\infty$ is very small, we at least get an estimate of size $|\lambda|^2$ for $d(\phi_\lambda(z), z)$.

4. Extension from S^1_∞ to $\mathbb{H}^2$ and other quasiconformal isotopy problems are discussed in [EaMc].

Barycenters and the baseball curve. To conclude we briefly discuss a natural construction due to Douady and Earle that extends any homeomorphism $f : S^{n-1}_\infty \rightarrow S^{n-1}_\infty$ to a smooth mapping $F : \mathbb{H}^n \rightarrow \mathbb{H}^n$.

Given a probability measure μ on the sphere with no atom of mass $1/2$ or more, there is a unique point $\beta(\mu) \in \mathbb{H}^n$ from which the measure appears to be balanced. This *barycenter* is characterized by the property that if one moves $\beta(\mu)$ to the center 0 in the ball model for hyperbolic space, then μ is transported to a measure whose Euclidean barycenter is also at the origin. The *barycenter extension* of f is defined by $F(p) = \beta(f_*(\mu_p))$, where μ_p is the visual measure on S_∞^{n-1} as seen from p.

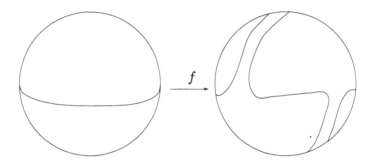

Figure B.2. The equator is sent to a baseball curve.

Since the construction only uses the elements of hyperbolic geometry, it is natural. That is, for any pair of hyperbolic isometries g and h, the barycenter extension of $g \circ f \circ h$ is $g \circ F \circ h$.

In dimension $n = 2$, the extended mapping F is always a diffeomorphism of the hyperbolic plane. Unfortunately this property fails to hold in dimension $n \geq 3$.

A counterexample is sketched in Figure B.2. Let $f : S_\infty^2 \to S_\infty^2$ be a smooth area-preserving map that sends the equator to a curve shaped something like the threads on a baseball. This curve divides the sphere into two interlocking congruent disks. The map f fixes the north and south poles, but it sends most of the northern hemisphere into the southern hemisphere, and vice-versa.

We can arrange that $A \circ R \circ f = f \circ A \circ R$, where A is the antipodal map and R is a $90°$ rotation fixing the poles. Then by naturality, F maps the geodesic γ joining the poles to itself with degree one, and F fixes the center 0 of the sphere. But as p moves north along γ from 0, the mass of $f_*(\mu_p)$ in the southern hemisphere increases, so

$F(p)$ initially moves south. Thus $F|\gamma$ is not a homeomorphism.

On the other hand, one can show that F is a quasi-isometry whenever the dilatation $K(f)$ is close enough to one. Thus another method for extending quasiconformal maps on the sphere is to first factor them into maps of small distortion, extend each of these by the barycenter construction, and then take the composition. As the factorization becomes finer and finer, this alternate construction converges to the visual isotopy extension presented above.

B.5 Almost isometries

The results of the preceding section allow one to construct quasi-isometries of hyperbolic space with explicit estimates on the pointwise distortion. (Compare Theorem 2.11.) To exploit the fact that a diffeomorphism has quasi-isometry constant close to 1, we need a stability result showing such a mapping is globally close to an isometry.

Theorem B.24 (Almost isometries) *Let $K \subset U$ be be a connected compact subset of an open set in $\mathbb{H}^n$, and let $\phi : U \to \mathbb{H}^n$ be a $(1 + \epsilon)$-quasi-isometric embedding. Then there is an isometry ι of $\mathbb{H}^n$ such that*

$$\sup_K d(\iota \circ \phi(x), x) \leq C(K, U)\epsilon.$$

To prove this result we use the fact that distances control the location of points.

Lemma B.25 *Let $\{v_i, i = 0, \dots, n\}$ be the vertices of a regular n-simplex with edge length $r > 0$ in hyperbolic space $\mathbb{H}^n$. Then for any $p, q \in \mathbb{H}^n$, if $d(p, v_0) \leq R$ and $|d(p, v_i) - d(q, v_i)| \leq \epsilon$ for all i, then $d(p, q) \leq C(n, r, R)\epsilon$.*

Proof. Consider the mapping $D : \mathbb{H}^n \to \mathbb{R}^{n+1}$ given by $D(p) = (d(p, v_0), \dots, d(p, v_n))$. It is easy to see D is an injective, proper mapping, using the fact that an n-simplex is determined up to isometry by the lengths of its edges. (In fact the last n coordinates of $D(p)$ determine p up to two possible choices, since they determine the simplex $(p, v_1, \ldots, v_n)$ up to isometry. The first coordinate distinguishes the two possibilities for p.)

Except at the vertices v_i, the mapping D is a smooth embedding. Indeed, the gradient of $-d(p, v_i)$ is in the direction of the geodesic from p to v_i, and these directions span the tangent space at p, so long as p is not a vertex. As for the vertices, at $p = v_i$ the coordinates $d(p, v_j)$ with $i \neq j$ provide a local diffeomorphism to $\mathbb{R}^n$.

It follows that the function

$$\frac{d(p, q)}{\sup_i |d(p, v_i) - d(q, v_i)|}$$

is bounded above on $K \times K - \{\text{the diagonal}\}$ for any compact set $K \subset \mathbb{H}^n$. The Lemma is an immediate consequence.

∎

Proof of Theorem B.24 (Almost isometries). First we treat the case where K and U are balls, say $K = \overline{B(p, R)} \subset U = B(p, S)$.

Suppose $\phi_k : U \to \mathbb{H}^n$ is a sequence of quasi-isometric embeddings, normalized so $\phi_k(p) = p$, and such that $L(\phi_k) \to 1$. Then $\langle \phi_k \rangle$ is equicontinuous, and ϕ_k is nearly volume-preserving for $k \gg 0$. Any uniform limit of a subsequence is a volume-preserving conformal map, hence an isometry. Thus when ϵ is small, a $(1 + \epsilon)$-quasi-isometry on U is uniformly close to an isometry on K.

It follows that the convex hull of $\phi(K)$ is contained in $\phi(U)$ whenever ϵ is sufficiently small. Since ϕ^{-1} is also a $(1 + \epsilon)$-quasi-isometry, we may apply ϕ^{-1} to the geodesic joining $\phi(p)$ to $\phi(q)$ to conclude that

$$\frac{1}{1 + \epsilon} d(p, q) \;\leq\; d(\phi(p), \phi(q)) \;\leq\; (1 + \epsilon) d(p, q)$$

for any $p, q \in K$.

Now let $(v_0, \ldots, v_n)$ be a regular simplex contained in K with edge length $R/2$. Then ϕ changes the lengths of the edges of the simplex by an additive factor which is $O(\epsilon)$, so there is another regular $R/2$-simplex $(w_0, \ldots, w_n)$ with $d(w_i, \phi(v_i)) \leq O(\epsilon)$. Let ι be the unique hyperbolic isometry such that $\iota(w_i) = v_i$. Then for any point $x \in K$, and any i, the distances $d(x, v_i)$ and $d(\iota \circ \phi(x), v_i)$ differ by an additive amount which is $O(\epsilon)$. By the preceding Lemma, this implies $d(x, \iota \circ \phi(x)) = O(\epsilon)$, completing the proof when K and U are concentric balls.

To treat the case of general (K, U), cover K with a finite collection of balls $B(p_i, R_i)$ such that $B(p_i, 2R_i) \subset U$. Then given ϕ we have isometries ι_i such that $d(\iota_i \circ \phi(x), x) = O(\epsilon)$ on $B(p_i, R_i)$. Whenever two balls overlap, we can conclude that ι_i and ι_j nearly agree on the overlap, so $d(\iota_i \circ \phi(x), \iota_j \circ \phi(x)) = O(\epsilon)$ for all $x \in K$. Since K is connected, any two balls are joined by a chain of overlapping balls. The total number of balls is finite, so $d(\iota_1 \circ \phi(x), x) = O(\epsilon)$ for all $x \in K$.

■

Stability results of this type for Euclidean space appear in [John]. More recent work on stability of quasi-isometries and quasiconformal maps is surveyed in [Res].

B.6   Points of differentiability

In this section we show that if $Mv \to 0$ rapidly along a geodesic in $\mathbb{H}^n$ terminating at $p \in S_\infty^{n-1}$, then v is *differentiable* at p. An integrated form of this result asserts that certain quasiconformal isotopies are differentiable, indeed nearly conformal, at p.

For convenience, we express these result in terms of the upper half-space model, where $\mathbb{H}^n = \{y = (y_1, \dots, y_n) \ : \ y_n > 0\}$, and $\mathbb{R}_\infty^{n-1} = \{y \ : \ y_n = 0\}$. Let $\gamma(s) = (0, \dots, 0, e^{-s})$ be a geodesic ray in $\mathbb{H}^n$, parameterized by arc length, converging to the origin $y = 0$ in $\mathbb{R}_\infty^{n-1}$ as $s \to \infty$.

A conformal vector field w is *linear* if $w(y) \in T_y \mathbb{R}_\infty^{n-1} \cong \mathbb{R}^{n-1}$ depends linearly on $y \in \mathbb{R}_\infty^{n-1}$; equivalently, if $w(0) = w(\infty) = 0$.

A vector field v is $C^{1+\alpha}$-*conformal* at a point $p \in \mathbb{R}_\infty^{n-1}$ if for all t sufficiently small,

$$v(p + t) = v(p) + v'(p) \cdot t + O(|t|^{1+\alpha}),$$

and $t \mapsto v'(p) \cdot t$ is a conformal vector field.

Theorem B.26 (Pointwise conformal vector field) *Let v be a vector field on $S_\infty^{n-1} = \mathbb{R}_\infty^{n-1} \cup \{\infty\}$. Suppose*

$$Mv(\gamma(s)) \le e^{-\alpha s}$$

for some $\alpha > 0$ and for all $s \geq 0$. Then $v(y)$ is $C^{1+\alpha}$-conformal at $y = 0$. Moreover, if $v(0) = v(\infty) = 0$, then

$$|v(y) - v'(0) \cdot y| \leq C(n, \alpha)|y|^{1+\alpha}$$

when $|y| \leq 1$.

Proof. After correcting by a conformal vector field, we may assume $v(0) = v(\infty) = 0$. For each $j \geq 0$, let $p_j = \gamma(j)$. From the hypothesis on $Mv(p_j)$ we find there is a linear conformal vector field w_j such that

$$\|v - w_j\|_\infty(p_j) = O(e^{-j\alpha}).$$

The linear vector field w_j is specified by a linear map $A_j : \mathbb{R}^{n-1} \to \mathbb{R}^{n-1}$.

Consider the annulus $W_j = \{y : e^{-j} > |y| > e^{-j-1}\}$. The visual metric on W_j as seen from either p_j or p_{j+1} is approximately equal to the Euclidean metric divided by $|y|$. Thus

$$|v(y) - w_j(y)| = O(e^{-j\alpha}|y|) = O(|y|^{1+\alpha})$$

on W_j, and therefore

$$\frac{|A_{j+1}(y) - A_j(y)|}{|y|} = O(|y|^\alpha).$$

Consequently $\|A_j - A_{j+1}\| = O(e^{-j\alpha})$ (using the usual norm on linear maps), so $A = \lim A_j$ exists and $\|A_j - A\| = O(e^{-j\alpha}/(1 - e^{-\alpha}))$.

Let w be the linear vector field corresponding to the linear map $A(y)$. If $|y| < 1$, then $y \in W_j$ for some j, and we have $|v(y) - w(y)| \leq C(n, \alpha)|y|^{1+\alpha}$. Since $v(0) = w(0) = 0$, we have $v'(0) = w'(0)$ and thus $w(y) = v'(0) \cdot y$.

∎

Since we will only use the integrated form of this assertion in conjunction with the Beltrami isotopy, we state it for the slightly simpler specific case of the Riemann sphere.

Definition. A map $\phi : \widehat{\mathbb{C}} \to \widehat{\mathbb{C}}$ is $C^{1+\alpha}$-*conformal* at $z \in \mathbb{C}$ if the complex derivative $\phi'(z)$ exists, and

$$\phi(z + t) = \phi(z) + \phi'(z) \cdot t + O(|t|^{1+\alpha})$$

for all t sufficiently small.

Theorem B.27 (Pointwise conformal map) *Let $\phi : \hat{\mathbb{C}} \times [0,1] \to$ $\hat{\mathbb{C}}$ be an isotopy generated by a continuous, time-dependent vector field v, such that $v_t(0) = v_t(1) = 0$. Suppose the complex derivative $v_t'(0)$ exists and for all t,*

$$|v_t(z) - v_t'(0)z)| \leq C|z|^{1+\alpha}$$

when $|z| \leq 1$. Then $\phi_1(z)$ is $C^{1+\alpha}$-conformal at $z = 0$, and

$$|\phi_1(z) - \phi_1'(0)z| \leq C'|z|^{1+\alpha}$$

when $|z| < \eta$, where C' and $\eta > 0$ depend only on C and α.

Proof. Since C is independent of t, $v_t(z)/z \to v_t'(0)$ uniformly in t, and thus $v_t'(0)$ is continuous. Also $|v_t'(0)| \leq C$ since $v_t(1) = 0$. Therefore when $|z| < 1$, we have

$$|v_t(z)| \leq |v_t'(0)||z| + C|z|^{1+\alpha} \leq 2C|z|.$$

By assumption, $d\phi_t(z)/dt = v_t(\phi_t(z))$. Thus the inequality above bounds the rate of change of $\log|\phi_t(z)|$, and we find that

$$|\phi_t(z)| \leq e^{2C}|z|$$

for all t, when $|z| < \eta = e^{-2C}$.

Let

$$\lambda_t = \exp\left(\int_0^t v_s'(0) \, ds\right),$$

and let $A_t(z) = \lambda_t z$ be the integral of the conformal vector field $v_t'(0)z$. That is, the time-dependent linear map $A_t(z)$ is the solution to the differential equation $A_0(z) = z$ and $dA_t(z)/dt = v_t'(0)A_t(z)$.

Let $\delta_t(z) = \phi_t(z) - A_t(z)$. Then for all $z < \eta$, we have

$$\begin{aligned}
\left|\frac{d\delta_t(z)}{dt}\right| &= \left|\frac{d\phi_t(z)}{dt} - \frac{dA_t(z)}{dt}\right| = |v_t(\phi_t(z)) - v_t'(0)A_t(z)| \\
&= |v_t(\phi_t(z)) - v_t'(0)\phi_t(z) + v_t'(0)(\phi_t(z) - A_t(z))| \\
&\leq C|\phi_t(z)|^{1+\alpha} + C|\phi_t(z) - A_t(z)| \\
&\leq C(e^{2C(1+\alpha)}|z|^{1+\alpha} + |\delta_t(z)|) \\
&= O(|z|^{1+\alpha} + |\delta_t(z)|).
\end{aligned}$$

Since $\delta_0(z) = 0$, this differential inequality implies $\delta_1(z) \leq C'|z|^{1+\alpha}$, where C' depends only on C and α. In particular, $|\phi_1(z) - A_1(z)| \leq C'|z|^{1+\alpha}$, so $\phi_1(z)$ is $C^{1+\alpha}$-conformal at the origin with $\phi_1'(0) = A_1'(0) = \exp \int_0^1 v_s'(0) \, ds$.

∎

B.7 Example: stretching a geodesic.

The vector field $V = \mathrm{ex}(v)$, its strain SV and the flow it generates can be computed explicitly in simple examples.

Consider the quasiconformal flow on the Riemann sphere given by $\phi_t(z) = |z|^{\exp(t)} z/|z|$. In logarithmic polar coordinates (ρ, θ) where $z = \exp(\rho + i\theta)$, we have $\phi_t(\rho, \theta) = (e^t \rho, \theta)$. The infinitesimal generator of this flow is the time-independent vector field

$$v = \rho \frac{\partial}{\partial \rho} = z \log |z| \frac{\partial}{\partial z};$$

and its conformal strain is given by

$$\mu = \bar{\partial} v = \frac{z}{2\bar{z}} \frac{d\bar{z}}{dz}.$$

Let $V = \mathrm{ex}(v)$. By integration, we obtain a volume preserving flow $\Phi_t : \mathbb{H}^3 \to \mathbb{H}^3$ extending ϕ_t and integrating V. By symmetry, Φ_t stabilizes the geodesic γ joining 0 to ∞ as well as the plane perpendicular to γ meeting $\hat{\mathbb{C}}$ in the unit circle.

Any point p in $\mathbb{H}^3$ lies in a hyperbolic plane $H(p)$ perpendicular to γ. Introduce cylindrical coordinates (r, θ, h) on $\mathbb{H}^3$, where (r, θ) are hyperbolic polar coordinates in $H(p)$, and $h(p)$ is the distance from $H(p)$ to the equatorial plane. (A ray perpendicular to γ travels distance $r(p)$ before hitting p and then terminates on $\hat{\mathbb{C}}$ at a point with argument θ.)

In these coordinates the hyperbolic metric becomes

$$ds^2 = dr^2 + \sinh^2(r) d\theta^2 + \cosh^2(r) dh^2$$

(using e.g. [Bea, §7.2 and §7.20]). The volume of the cylinder

$$C(r, h) = [0, r] \times [0, 2\pi] \times [0, h]$$

is therefore $\pi h \sinh^2(r)$.

By symmetry, $\Phi_t(r, h, \theta) = (r_t, e^t h, \theta)$. Since Φ_t is volume preserving, and $\Phi_t(C(r, h)) = C(r_t, e^t h)$, at $t = 0$ we have

$$0 = \frac{d \operatorname{vol} C(r_t, e^t h)}{dt} = \pi h \left(\sinh^2(r) + 2 \sinh(r) \cosh(r) \frac{dr_t}{dt} \right);$$

therefore $dr_t/dt = -\tanh(r)/2$ and

$$V = \frac{d\Phi_t}{dt}\bigg|_{t=0} = -\frac{\tanh(r)}{2} \frac{\partial}{\partial r} + h \frac{\partial}{\partial h}.$$

A computation, using formula (A.5), shows the strain tensor SV is diagonal in the cylindrical coordinates (r, θ, h), with diagonal entries

$$\left(\frac{-1}{2\cosh^2(r)}, \ -\frac{1}{2}, \ \frac{1}{2} + \frac{1}{2\cosh^2(r)} \right).$$

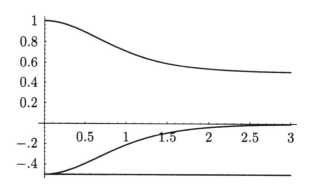

Figure B.3. Strain eigenvalues.

The behavior of the eigenvalues as a function of r is graphed in Figure B.3. At $r = 0$ the eigenvalues are $(-1/2, -1/2, 1)$ with maximum strain in the h direction; as $r \to \infty$ the radial strain tends to zero and SV tends to the tangential strain μ, with eigenvalues $(0, -1/2, 1/2)$.

In this example, Φ_t is a linear stretch with factor e^t on the geodesic γ. For any hyperbolic isometry $g_0(z) = \lambda_0 z$, we have $g_t \circ \phi_t = \phi_t \circ g_0$, where $g_t(z) = \lambda^{\exp(t)} z$. If $|\lambda_0| > 1$, then the group Γ_t generated by g_t is discrete, and the manifold $M_t = \mathbb{H}^3/\Gamma_t$ is a solid torus with core geodesic of length $e^t \log|\lambda_0|$. The map Φ_t descends to a volume-preserving quasi-isometry from M_0 to M_t. The quasi-isometry constant of Φ_t is exactly e^t; its distortion is maximum on the geodesic.

Notes and references. The theory of visual extension of vector fields from S^{n-1} to $\mathbb{H}^n$ is developed in [Ah2], [Th1, Chapter 11], [Rei3] and [Gai]. Ahlfors studied general strain-harmonic vector fields on hyperbolic space and showed, with suitable regularity at infinity, that they can be recovered from their boundary values by visual extension. The idea that uniqueness of the visual extension is related to Frobenius reciprocity appears in [Earle]. For background in representation theory one may refer to [Kn] and [BD]; decompositions of representations over subgroups are discussed in [Zel, §XVIII]. The applications of the visual extension to the generation of flows and quasi-isometries, given in Theorem B.21 and Corollary B.23, were established by Reimann [Rei3], completing the development begun by Ahlfors and Thurston.

Some papers implicitly appeal to compactness of K-quasiconformal maps, but using a measure of the dilatation different from the one we use here (such as the linear dilatation). To our knowledge these compactness results are not available in the published literature.

Bibliography

[AK]   S. B. Agard and J. Kelingos. On parameteric represen-
          tation for quasisymmetric functions. *Comm. Math. Helv.*
          44(1969), 446–456.

[Ah1]   L. Ahlfors. Conditions for quasiconformal deformations in
          several variables. In *Contributions to Analysis*, pages 19–25.
          Academic Press, 1974.

[Ah2]   L. Ahlfors. Invariant operators and integral representations
          in hyperbolic space. *Math. Scan.* **36**(1975), 27–43.

[Ah3]   L. Ahlfors. Quasiconformal deformations and mappings in
          R^n. *J. d'Analyse Math.* **30**(1976), 74–97.

[Ah4]   L. Ahlfors. A somewhat new approach to quasiconformal
          mappings in R^n. In *Complex Analysis*. Springer-Verlag Lec-
          ture Notes in Math. 599, 1977.

[Ah5]   L. Ahlfors. Some remarks on Kleinian groups. In *Collected
          Papers*, volume 2, pages 316–319. Birkhauser, 1982.

[AB]   L. Ahlfors and L. Bers. Riemann's mapping theorem for
          variable metrics. *Annals of Math.* **72**(1960), 385–404.

[Ast]   K. Astala. Area distortion of quasiconformal mappings.
          Acta Math. **173**(1994), 37–60.

[Bea]   A. Beardon. *The Geometry of Discrete Groups*. Springer-
          Verlag, 1983.

[BP]   R. Benedetti and C. Petronio. *Lectures on Hyperbolic Ge-
          ometry*. Springer-Verlag, 1992.

[Bers1]   L. Bers. On boundaries of Teichmüller spaces and on
          kleinian groups: I. *Annals of Math.* **91**(1970), 570–600.

[Bers2]   L. Bers. An extremal problem for quasiconformal maps and
          a theorem by Thurston. *Acta Math.* **141**(1978), 73–98.

241

[Bers3]   L. Bers. On iterates of hyperbolic transformations of Te-
          ichmüller space. *Amer. J. Math.* **105**(1983), 1–11.

[BJ]   C. J. Bishop and P. W. Jones. Hausdorff dimension and
          Kleinian groups. *Preprint.*

[Bon]   F. Bonahon. Bouts des variétés hyperboliques de dimension
          3. *Annals of Math.* **124**(1986), 71–158.

[BD]   T. Bröcker and T. tom Dieck. *Representations of Compact
          Lie Groups.* Springer-Verlag, 1985.

[Can]   R. D. Canary. A covering theorem for hyperbolic 3-
          manifolds and its applications. *To appear, Topology.*

[CEG]   R. D. Canary, D. B. A. Epstein, and P. Green. Notes on
          notes of Thurston. In *Analytical and Geometric Aspects of
          Hyperbolic Space*, pages 3–92. Cambridge University Press,
          1987.

[CaTh]   J. W. Cannon and W. P. Thurston. Group invariant Peano
          curves. *Preprint.*

[Car]   C. Carathéodory. *Conformal Representation.* Cambridge
          University Press, 1952.

[CG]   L. Carleson and T. Gamelin. *Complex Dynamics.* Springer-
          Verlag, 1993.

[CoTr]   P. Coullet and C. Tresser. Itération d'endomorphismes
          et groupe de renormalisation. *J. de Physique Colloque C*
          539(1978), C5–25.

[Cvi]   P. Cvitanović. *Universality in Chaos.* Adam Hilger Ltd,
          1984.

[dF]   E. de Faria. Asymptotic rigidity of scaling ratios for critical
          circle mappings. *Preprint.*

[dFdM]   E. de Faria and W. de Melo. Rigidity of critical circle map-
          pings. *In preparation.*

[Dou1] A. Douady. Chirurgie sur les applications holomorphes. In *Proceedings of the International Congress of Mathematicians*, pages 724–738. Amer. Math. Soc., 1986.

[Dou2] A. Douady. Disques de Siegel et anneaux de Herman. *Astérisque* **152-153**(1987), 151–172.

[Dou3] A. Douady. Does a Julia set depend continuously on the polynomial? In R. Devaney, editor, *Complex Analytic Dynamics*. AMS Proc. Symp. Appl. Math. 49, 1994.

[DH] A. Douady and J. Hubbard. On the dynamics of polynomial-like mappings. *Ann. Sci. Éc. Norm. Sup.* **18**(1985), 287–344.

[EaMc] C. Earle and C. McMullen. Quasiconformal isotopies. In *Holomorphic Functions and Moduli I*, pages 143–154. Springer-Verlag: MSRI publications volume 10, 1988.

[Earle] C. J. Earle. Conformally natural extension of vector fields from S^{n-1} to B^n. *Proc. Amer. Math. Soc.* **102**(1988), 145–149.

[EpM] D. B. A. Epstein and A. Marden. Convex hulls in hyperbolic space, a theorem of Sullivan, and measured pleated surfaces. In *Analytical and Geometric Aspects of Hyperbolic Space*, pages 113–254. Cambridge University Press, 1987.

[FLP] A. Fathi, F. Laudenbach, and V. Poénaru. Travaux de Thurston sur les surfaces. *Astérisque* **66-67**(1979).

[Feig] M. Feigenbaum. Quantitative universality for a class of non-linear transformations. *J. Stat. Phys.* **19-1**(1978), 25–52.

[Gai] P.-Y. Gaillard. Transformation de Poisson de formes differentielles. Le cas de l'espace hyperbolique. *Comment. Math. Helv.* **61**(1986), 581–616.

[Gard] F. Gardiner. *Teichmüller Theory and Quadratic Differentials*. Wiley Interscience, 1987.

[GV]   F. W. Gehring and J. Väisälä. Hausdorff dimension and quasiconformal mappings. *J. London Math. Soc.* **6**(1973), 504–512.

[Gol]   W. M. Goldman. The symplectic nature of fundamental groups of surfaces. *Adv. Math.* **54**(1984), 200–225.

[GH]   P. Griffiths and J. Harris. *Principles of Algebraic Geometry.* Wiley Interscience, 1978.

[Gun1]   R. Gunning. *Lectures on Vector Bundles over Riemann Surfaces.* Princeton University Press, 1967.

[Gun2]   R. Gunning. *Introduction to Holomorphic Functions of Several Complex Variables.* Wadsworth & Brooks/Cole, 1990.

[Haus]   F. Hausdorff. *Set Theory.* Chelsea, 1957.

[Hel]   S. Helgason. *Groups and Geometric Analysis.* Academic Press, 1984.

[Her]   M. Herman. Conjugaison quasi-symétrique des difféomorphismes du cercle et applications aux disques singuliers de Siegel. *Manuscript, 1986.*

[Hil]   E. Hille. *Lectures on Ordinary Differential Equations.* Addison-Wesley, 1969.

[HY]   J. G. Hocking and G. S. Young. *Topology.* Addison-Wesley, 1961.

[Ji]   Y. Jiang. Infinitely renormalizable quadratic Julia sets. *Preprint, 1993.*

[JH]   Y. Jiang and J. Hu. The Julia set of the Feigenbaum quadratic polynomial is locally connected. *Preprint, 1993.*

[John]   F. John. Rotation and strain. *Comm. Pure App. Math.* **14**(1961), 391–414.

[JN]   F. John and L. Nirenberg. On functions of bounded mean oscillation. *Comm. Pure App. Math.* **14**(1961), 415–426.

[Jor]   T. Jørgensen. Compact 3-manifolds of constant negative curvature fibering over the circle. *Ann. Math.* **106**(1977), 61–72.

[Ka]   J. Kahn. Teichmüller theory and renormalization. *In preparation.*

[KT]   S. Kerckhoff and W. Thurston. Non-continuity of the action of the modular group at Bers' boundary of Teichmüller space. *Invent. math.* **100**(1990), 25–48.

[Kn]   A. W. Knapp. *Representation Theory of Semisimple Groups.* Princeton University Press, 1986.

[LV]   O. Lehto and K. J. Virtanen. *Quasiconformal Mappings in the Plane.* Springer-Verlag, 1973.

[Lyu]   M. Lyubich. Geometry of quadratic polynomials: Moduli, rigidity, and local connectivity. *Stony Brook IMS Preprint 1993/9.*

[MSS]   R. Mañé, P. Sad, and D. Sullivan. On the dynamics of rational maps. *Ann. Sci. Éc. Norm. Sup.* **16**(1983), 193–217.

[Mc1]   C. McMullen. Iteration on Teichmüller space. *Invent. math.* **99**(1990), 425–454.

[Mc2]   C. McMullen. Cusps are dense. *Annals of Math.* **133**(1991), 217–247.

[Mc3]   C. McMullen. Rational maps and Kleinian groups. In *Proceedings of the International Congress of Mathematicians Kyoto 1990*, pages 889–900. Springer-Verlag, 1991.

[Mc4]   C. McMullen. *Complex Dynamics and Renormalization.* Annals of Math Studies 135, Princeton University Press, 1994.

[Mc5]   C. McMullen. The classification of conformal dynamical systems. In *Current Developments in Mathematics, 1995.* To appear, International Press, 1996.

[Mc6]   C. McMullen. Complex earthquakes and Teichmüller the-
        ory. *Preprint, 1996.*

[Mc7]   C. McMullen. Self-similarity of Siegel disks and the Haus-
        dorff dimension of Julia sets. *Preprint, 1995.*

[McS]   C. McMullen and D. Sullivan. Quasiconformal homeomor-
        phisms and dynamics III: The Teichmüller space of a holo-
        morphic dynamical system. *To appear, Adv. Math.*

[MeSt] W. de Melo and S. van Strien. *One-Dimensional Dynamics.*
        Springer-Verlag, 1993.

[Mil]   J. Milnor. Self-similarity and hairiness in the Mandelbrot
        set. In M. C. Tangora, editor, *Computers in Geometry and
        Topology*, pages 211–259. Lect. Notes Pure Appl. Math.,
        Dekker, 1989.

[Min]   Y. Minsky. On rigidity, limit sets, and end invariants of
        hyperbolic 3-manifolds. *J. Amer. Math. Soc* **7**(1994), 539–
        588.

[Mor]   J. Morgan. Group actions on trees and the compactification
        of the space of classes of $SO(n,1)$-representations. *Topology*
        25(1986), 1–33.

[Mos]   D. Mostow. *Strong rigidity of locally symmetric spaces.* An-
        nals of Math Studies 78, Princeton University Press, 1972.

[MMW] D. Mumford, C. McMullen, and D. Wright. Limit sets of
        free two-generator kleinian groups. *Preprint, 1990.*

[Nad]   S. B. Nadler. *Hyperspaces of Sets.* Marcel Dekker, Inc, 1978.

[NS]   T. Nowicki and S. van Strien. Polynomial maps with a Julia
        set of positive Lebesgue measure: Fibonacci maps. *Preprint,
        1993.*

[Otal] J.-P. Otal. *Le théorème d'hyperbolisation pour les variétés
        fibrées de dimension trois.* Astérisque, to appear.

[Rei1]   H. M. Reimann. Functions of bounded mean oscillation and quasiconformal mappings. *Comment. Math. Helv.* **49**(1974), 260–276.

[Rei2]   H. M. Reimann. Ordinary differential equations and quasi-conformal mappings. *Invent. math* **33**(1976), 247–270.

[Rei3]   H. M. Reimann. Invariant extension of quasiconformal deformations. *Ann. Acad. Sci. Fen.* **10**(1985), 477–492.

[Res]   Yu. G. Reshetnyak. *Space Mappings with Bounded Distortion*. AMS Translations vol. 73, 1989.

[Sal]   A. Salli. On the Minkowski dimension of strongly porous fractal sets in $\mathbb{R}^n$. *Proc. London Math. Soc.* **62**(1991), 353–372.

[Sh]   H. Shiga. On Teichmüller spaces and modular transformations. *J. Math. Kyoto Univ.* **25**(1985), 619–626.

[St]   J. Stallings. On fibering certain 3-manifolds. In *Topology of 3-Manifolds*, pages 95–100. Prentice Hall, 1962.

[Stein1] E. M. Stein. Singular integrals, harmonic functions, and differentiability properties of functions of several variables. In *Singular Integrals*, pages 316–335. AMS Symposia in Pure Math. X, 1967.

[Stein2] E. M. Stein. *Singular Integrals and Differentiability Properties of Functions*. Princeton Univeristy Press, 1970.

[SZ]   E. M. Stein and A. Zygmund. Boundedness of translation invariant operators on Hölder spaces and L^p-spaces. *Annals of Math.* **85**(1967), 337–349.

[Sul1]   D. Sullivan. Travaux de Thurston sur les groupes quasi-fuchsiens et sur les variétés hyperboliques de dimension 3 fibrées sur S^1. *Sem. Bourbaki* **554**(1979/80).

[Sul2]   D. Sullivan. Growth of positive harmonic functions and Kleinian group limit sets of zero planar measure and Hausdorff dimension two. In *Geometry at Utrecht*, pages 127–144. Springer-Verlag Lecture Notes in Math. 894, 1981.

[Sul3]   D. Sullivan. On the ergodic theory at infinity of an arbitrary discrete group of hyperbolic motions. In I. Kra and B. Maskit, editors, *Riemann Surfaces and Related Topics: Proceedings of the 1978 Stony Brook Conference.* Annals of Math. Studies 97, Princeton, 1981.

[Sul4]   D. Sullivan. Quasiconformal homeomorphisms and dynamics I: Solution of the Fatou-Julia problem on wandering domains. *Annals of Math.* **122**(1985), 401–418.

[Sul5]   D. Sullivan. Bounds, quadratic differentials and renormalization conjectures. In F. Browder, editor, *Mathematics into the Twenty-first Century: 1988 Centennial Symposium, August 8-12*, pages 417–466. Amer. Math. Soc., 1992.

[Sw]   G. Świątek. Hyperbolicity is dense in the real quadratic family. *Stony Brook IMS Preprint 1992/10.*

[Tan]   Tan L. Similarity between Mandelbrot set and Julia sets. *Commun. Math. Phys.* **134**(1990), 587–617.

[Th1]   W. P. Thurston. *Geometry and Topology of Three-Manifolds.* Princeton lecture notes, 1979.

[Th2]   W. P. Thurston. Hyperbolic structures on 3-manifolds I: Deformations of acylindrical manifolds. *Annals of Math.* **124**(1986), 203–246.

[Th3]   W. P. Thurston. On the geometry and dynamics of diffeomorphisms of surfaces. *Bull. Amer. Math. Soc.* **19**(1988), 417–432.

[Th4]   W. P. Thurston. *Three-Dimensional Geometry and Topology.* University of Minnesota preprint, 1990.

[Th5]   W. P. Thurston. Hyperbolic structures on 3-manifolds II: Surface groups and 3-manifolds which fiber over the circle. *Preprint.*

[Vai]   J. Väisälä. *Lectures on n-Dimensional Quasiconformal Mappings.* Springer-Verlag Lecture Notes 229, 1971.

[Vuo1]   M. Vuorinen. *Conformal Geometry and Quasiregular Mappings*. Springer-Verlag Lecture Notes 1319, 1988.

[Vuo2]   M. Vuorinen, editor. *Quasiconformal Space Mappings*. Springer-Verlag Lecture Notes 1508, 1992.

[Wr]   D. Wright. The shape of the boundary of Maskit's embedding of the Teichmüller space of once-punctured tori. *Preprint.*

[Zel]   D. P. Zelobenko. *Compact Lie Groups and their Representations*. AMS Translations of Math. Monographs, vol. 40, 1973.

[Zie]   W. P. Ziemer. *Weakly Differentiable Functions*. Springer-Verlag, 1989.

Index